Human Factors of Outer Space Production

AAAS Selected Symposia Series

Published by Westview Press, Inc.
5500 Central Avenue, Boulder, Colorado

for the

American Association for the Advancement of Science
1776 Massachusetts Avenue, N.W., Washington, D.C.

Human Factors of Outer Space Production

Edited by
T. Stephen Cheston and David L. Winter

AAAS Selected Symposium **50**

AAAS Selected Symposia Series

This book is based on a symposium which was held at the 1979 AAAS National Annual Meeting in Houston, Texas, January 3-8. The symposium was sponsored by the American Institute of Aeronautics and Astronautics.

All rights reserved. No part of this publication may be reproduced or transmitted in any form or by any means, electronic or mechanical, including photocopy, recording, or any information storage and retrieval system, without permission in writing from the publisher.

Copyright © 1980 by the American Association for the Advancement of Science

Published in 1980 in the United States of America by
  Westview Press, Inc.
  5500 Central Avenue
  Boulder, Colorado 80301
  Frederick A. Praeger, Publisher

Library of Congress Cataloging in Publication Data
Main entry under title:
Human factors of outer space production.
   (AAAS selected symposia series ; 50)
   Includes bibliographies.
   1. Space colonies. 2. Space flight--Psychological aspects. 3. Space stations--Industrial applications. I. Cheston, T. Stephen. II. Winter, David L. III. Series: American Association for the Advancement of Science. AAAS selected symposia series ; 50.
TL795.7.H85   629.44'2   79-24648
ISBN 0-89158-789-6

Printed and bound in the United States of America

About the Book

The missions of the early space age--when a relatively few, very highly trained, physically fit male pilot/astronauts operated for short times--will be supplemented in the future by missions where large numbers of nonpilot/astronaut men and women will work in orbit for long periods of time on research and industry-related tasks. The lengthening and changing complexity of space operations requires that the psychosocial, habitat design, food systems, and economic aspects of humans working in space be reviewed carefully. In this volume, an interdisciplinary group of experts addresses these aspects of space work and delineates avenues for future research.

About the Series

The *AAAS Selected Symposia Series* was begun in 1977 to provide a means for more permanently recording and more widely disseminating some of the valuable material which is discussed at the AAAS Annual National Meetings. The volumes in this *Series* are based on symposia held at the Meetings which address topics of current and continuing significance, both within and among the sciences, and in the areas in which science and technology impact on public policy. The *Series* format is designed to provide for rapid dissemination of information, so the papers are not typeset but are reproduced directly from the camera-copy submitted by the authors, without copy editing. The papers are organized and edited by the symposium arrangers who then become the editors of the various volumes. Most papers published in this *Series* are original contributions which have not been previously published, although in some cases additional papers from other sources have been added by an editor to provide a more comprehensive view of a particular topic. Symposia may be reports of new research or reviews of established work, particularly work of an interdisciplinary nature, since the AAAS Annual Meetings typically embrace the full range of the sciences and their societal implications.

WILLIAM D. CAREY
Executive Officer
American Association for
the Advancement of Science

Contents

Figures and Tables xi

About the Editors and Authors xv

Preface xix

1 Psychological Considerations in Future
   Space Missions--*Robert L. Helmreich, John
   A. Wilhelm and Thomas E. Runge* 1

   Introduction   1
   Factors in Personnel Require-
      ments   2
      *New Missions in Space, 2; Changes
      in Crew Composition, 3; Changes in
      the Costs and Rewards of Space-
      flight, 3; Crowding and Privacy in
      the Space Habitat, 4; Leisure
      Aboard Spacecraft, 8; Psychological
      Selection of Crews, 9; Authority
      Structure, 14; Sex, 16; Mission
      Duration, 17; The Need for Research,
      18*

   References and Notes   19

2 Future Directions for Selecting Per-
   sonnel--*Kirmach Natani* 25

   Summary   25
   Introduction   27
   Antarctic Research: An Un-
      tapped Resource   29

Contents

      Crew Selection Studies, 29;
      Antarctica as a Field Laboratory,
      30; Hypoxic Stress and New Learning, 32; Task Sensitization with
      Lateralization Techniques, 33;
      Cerebral Lateralization and
      Adaptive Ability, 35
    Psychological Adaptability
      and Competence   37
      Selecting for Competence, 39;
      Observations from the Field,
      40; Continuing Reassessment,
      42; Costs and Rewards of Isolated Duty, 43
    Acknowledgements   47
    References and Notes   48

3   Well-Being and Privacy in Space:
    Anticipating Conflicts of Interest--
    Joan E. Sieber 65

    Introduction   65
    The Relativity of Privacy   67
    Privacy, Stress and Performance   70
    Anticipating Some Conflicts
      of Interest with Need for
      Privacy   74
    References and Notes   77

4   Habitat Requirements, Design and Options--
    Richard L. Kline 79

    Abstract   79
    Discussion   79
      Space Habitation Evolution, 79;
      Man's Function in Space and System Concepts, 83; Special Habitation Requirements, 89
    Conclusions and Recommendations   94
    References and Notes   96

5   Man-Machine Design for Spaceflight--
    Allen J. Louviere and John T. Jackson 97

    Introduction   97

Anthropometric Changes in
  Weightlessness   98
  *Concepts of Percentiles in Design,
  99; Posture, 101; Trends, 105;
  Size Differences Between Sexes,
  106; Variation Between Whites and
  Blacks, 106; Variation Among Na-
  tionalities, 107; Variation Among
  International Crewmen, 107; A
  Basic Element of Shuttle Work-
  station Design, 109; Remote Manip-
  ulator System Control Station
  Design, 112*

6 Nutritional Criteria for Closed-Loop
   Space Food Systems--*Paul C. Rambaut* 113

   Introduction   113
   Nutritional Requirements in
     Space   115
     *Fat, 117; Carbohydrates, 118; Amino
     Acids, 118; Macronutrient Elements,
     121; Micronutrient Elements, 124*
   Food Production Considerations   125
   References and Notes   130

7 Rationale for Evaluating a Closed
   Food Chain for Space Habitats--
   Michael Modell and Jack M. Spurlock 133

   Introduction   133
   Completely Closed Systems   135
   Rationale for Evaluating Closed
     Systems   139
   References and Notes   144

8 Problems of Food Technology in Space
   Habitats--*Marcus Karel* 147

   References and Notes   157

9 Use of Phytotrons in Assessing Environ-
   mental Requirements for Plants in Space
   Habitats--*C. David Raper, Jr., Terry A.
   Pollock and Judith Fey Thomas* 159

   Introduction   159

Changes in Plant Growth Attributable to Environment 161
Dynamic Crop Simulators 164
Cultural and Genetic Manipulations 165
Other Factors 166
Summary 166
References and Notes 167

10 Controlled-Environment Agricultural Systems for Large Space Habitats-- *John M. Phillips* 169

Background 169
Controlled-Environment Agriculture 171
Future Research 173
Terrestrial Applications and Benefits 174
Conclusions 176
References and Notes 176

11 Economic Factors of Outer Space Production-- *B. P. Miller* 179

Introduction 179
The Nature of Outer Space Production 179
Transportation 183
Future Opportunities for Outer Space Production 185
  Earth Resources Management Data, 185; Land Mobile Communications Services, 192; Research Product Areas, 198
Other Economic and Policy Issues 203
References and Notes 205

Figures and Tables

Chapter 4

Figure 1	Space habitation evolution	80
Figure 2	Man's functions in space	82
Figure 3	Open cherry picker	84
Figure 4	LEO observation platform	86
Figure 5	Closed cabin cherry picker	87
Figure 6	Manned orbit transfer vehicle	87
Figure 7	Cabin module for manned orbit transfer vehicle	88
Figure 8	Permanent large-scale space habitation	90
Figure 9	Habitation requirements	91
Figure 10	Responses to severe emergency on GEO sortie mission	92
Figure 11	Conclusions	95

Chapter 5

Figure 1	Neutral body positions of space	100
Figure 2	Geometry of the weightless body	102
Figure 3	Trends in individual growth	104

xii   Figures and Tables

Figure 4	Crew size variations: American, foreign and international	108
Figure 5	Foot restraint	109
Figure 6	RMS Control Station	110
Figure 7	Crew size and neutral body position criteria for RMS Control Station	111

Chapter 6

Figure 1	Food supplies nutritional needs through its nutrient constituents	114
Figure 2	Calorie expenditure rates during the Skylab missions	116
Figure 3	Change in urinary and fecal nitrogen as a function of Skylab flight duration	120
Figure 4	Daily calcium balance as a function of time during and after Skylab Earth Orbital Missions	122
Figure 5	Changes in urinary and fecal calcium as a function of time during and after Skylab Earth Orbital Missions	123
Figure 6	Alternative methods of producing food inflight	126
Figure 7	Concept for a Spacelab experiment with an intensive agriculture system	128

Chapter 7

Table 1	Inputs required to support a person in space	134
Figure 1	Example of a partially closed life support system	136
Figure 2	Inputs and outputs for a generalized closed-ecology life support system	138
Figure 3	Schematic of a completely closed system	140
Table 2	Typically chosen options for scenario components	141

Figure 4   Procedure for scenario formulation   142

Chapter 8

Table 1	Menu of a meal consumed on the way to the moon: fiction	148
Table 2	Menu of a typical meal consumed on the way to the moon: reality	148
Table 3	Anticipated systems for space feeding	149
Table 4	Where the food dollar goes	149
Table 5	Energy flow in U.S. food chain	150
Figure 1	Technological requirements for a range of hypothetical diets	152
Table 6	Some food processing operations	153
Table 7	Conversion of rice to food	153
Table 8	Human nutrition in an isolated environment – research needs	155
Table 9	Research and development tasks for food technology in space (process and equipment design)	156
Table 10	Research and development tasks for food technology in space (social and human factors)	156

Chapter 9

Table 1	Control limits in the North Carolina State University Phytotron	160
Table 2	Effects of photoperiod and temperature during pod-fill on growth of soybeans	162

Chapter 11

Figure 1	Current status of outer space production	182
Figure 2	Elements of an information and decision system for resources management	188

Table 1	Present value of benefits and costs of the LANDSAT operational system	191
Table 2	Annual benefits of LANDSAT operational system	191
Figure 3	Emergency Medical Service system model	194
Figure 4	Cost effectiveness of satellite communications	196
Figure 5	Materials processing in space--long-range plan	200
Figure 6	Satellite solar power system configuration	202

About the Editors and Authors

T. Stephen Cheston, *an historian by training, is associate dean of the graduate school at Georgetown University and vice-chairman of the Board of Trustees of the Universities Space Research Association. He has been involved in space-related activities since 1975, has established a working group to assess the impact of space on society from a social science and humanities perspective, and was founder of the Institute for the Social Science Study of Space (1978). He is co-editor of* Space Humanization Series, Vol. I *(with D. Webb; Institute for the Social Science Study of Space, 1979) and author of various articles on space social science.*

David L. Winter *is director of the Medical Research Administration at Sandoz, Inc., and the U.S. co-chairman of the joint US/USSR Working Group in Space Medicine and Biology. A neurophysiologist by training, he was formerly chief of the Division of Neuropsychiatry, Walter Reed Army Institute of Research, and was director for Life Sciences for NASA. He is the author of numerous publications in neurophysiology and space medicine.*

Robert L. Helmreich *is professor of psychology and chairman of the graduate program in social psychology at the University of Texas. He is chairman of the executive committee of the Society of Experimental Social Psychology and a fellow of the American Psychological Association. His area of specialization is social and personality psychology, and he has published on the topics of group behavior under stress and isolation and on achievement motivation. He is co-author of* Masculinity and Femininity: Their Psychological Dimensions, Antecedents and Correlates *(with Janet T. Spence; University of Texas Press, 1978).*

John T. Jackson, *a research psychologist/human factors engineer at the NASA-Johnson Space Center, is concerned with* evaluation of human capability to perform in space and with methods of analyzing crew performance data for future space missions. He has served as principal investigator or consultant on various Skylab experiments and was a member of the evaluation team for selecting crewmembers for Shuttle.

Marcus Karel is a professor of food engineering, Department of Nutrition and Food Science, Massachusetts Institute of Technology, and is a fellow of the U.S. and British Institutes of Food Technologists. He has been a consultant to the Johnson Space Center during the Apollo and Skylab missions and to various other scientific and industrial organizations. He has published a book and over 130 research papers on various aspects of food chemistry and engineering, and has received awards from the Institute of Food Technologists, the American Association of Agricultural Engineers, and the Dairy and Food Industries Supply Association.

Richard L. Kline is manager for civil space systems at Grumman Aerospace Corporation. He has been concerned with engineering activities related to the fields of thermodynamics, propulsion, design, cryogenics and the infrared. He worked on the NASA Lunar Module and Space Shuttle programs and is currently responsible for a number of Grumman's advanced projects, including space solar power systems, space transportation, and fabrication and large space structure demonstration projects.

Allen J. Louviere is chief of the Spacecraft Design Division, NASA-Johnson Space Center, which is responsible for the mechanical systems management of the Orbiter and for research on solar power satellites, Skylab reboost, and advanced design of large space structures. He worked on concept designs for Gemini, Apollo Lunar Module, and manned Venus and Mars missions and was responsible for crew station and habitability design for Skylab.

B. P. Miller is vice president for technology assessment, ECON, Inc., and directed the SEASAT economic assessment study of the uses, data requirements and economics of an oceanographic satellite system. He also directed the study of the economics of satellite communications and of processing biological materials in space for NASA and has written several articles concerning the economics of technology assessment. A former associate editor of the Journal of Spacecraft and Rockets, he is now a member of the International Activities Committee of the American Institute of Aeronautics and Astronautics.

About the Editors and Authors   xvii

Michael Modell *is associate professor of chemical engineering at Massachusetts Institute of Technology. A consultant to Arthur D. Little, Inc., and a member of the SAE/NASA Bioenvironmental Systems Study Group, his research interests include thermodynamics, heterogeneous catalysis, waste treatment and pollution control.*

Kirmach Natani, *a National Research Council resident research associate at the USAF School of Aerospace Medicine, is conducting computer-assisted neuropsychological assessments of aircrew personnel for flight duty, and he is a consultant to the Space Utilization Team at Georgetown University. He has also carried out research on the sleep psychophysiology of men wintering at the Amundsen-Scott South Pole Station. Among his numerous publications is "Ecopsychiatric Aspects of a First Human Space Colony" (with J. T. Shurley and R. A. Sengel) in* Space Manufacturing Facilities (Space Colonies) II *(J. Grey, ed.; American Institute of Aeronautics and Astronautics, 1977).*

John M. Phillips *is president and senior research associate at Arizona Research Associates, Inc. An agricultural biologist by training, his specific areas of specialization are educational and research activities in controlled-environment agriculture, solar energy (biomass) and space settlement and industrialization studies. He has worked on plant pathology and mineral nutrition in arid lands in the Near East and on hydroponic and drip irrigation projects in the United States. He has prepared a number of reports on these topics.*

Terry A. Pollock *is a graduate student in human nutrition at East Carolina University and has worked with the growth chambers system of the Phytotron at North Carolina State University. She has also worked at the Clayton Research Station on a project concerning environmental influences on the nutritional value of soybeans.*

Paul C. Rambaut *is manager of the biomedical research program of the Life Sciences Division at NASA; he was formerly deputy program manager for nutrition at the Food and Drug Administration and a member of the Nutrition Coordination Committee of the Department of Health, Education and Welfare. An associate fellow of the Aerospace Medical Association, he has numerous publications in his fields of interest, including "Calcium and Nitrogen Balance in Crewmembers of the 84-day Skylab IV Orbital Mission" (with C. S. Leach and G. D. Whedon;* Acta Astronautica, *in press) and "Prolonged Weightlessness and Bone Loss in Man"* (American Journal of Clinical Nutrition, *in press).*

C. David Raper, Jr., *is an associate professor in the department of soil science at North Carolina State University. He has published many papers dealing with plant growth in controlled environments and has participated in NASA workshops on space settlements and ecological life-support systems.*

Thomas E. Runge, *of the department of psychology at the University of Texas, is a specialist in leisure research and interpersonal relationships.*

Joan E. Sieber *is a professor of psychology at California State University (Hayward) and a senior visiting research scholar at the Kennedy Institute of Ethics, Georgetown University. She is director of an NSF project to investigate ethical issues in social science research, and she has carried out research for NASA on the effects of isolation and extended confinement on interpersonal relationships. She is author of* Anxiety, Learning and Instruction *(Hillsdale, NJ: Lawrence Erlbaum Associates, 1977) and is the editor of* Ethical Decision Making of Social Science Research *(NY: Cambridge, in press).*

Jack M. Spurlock *is director of the Chemical and Material Sciences Laboratory at the Georgia Institute of Technology and chairman of the committee of Spacecraft Environmental Control and Life Support Systems of the Society of Automotive Engineers. He is a fellow of the Royal Society of Health and the American Institute of Chemists, and an associate fellow of the Aerospace Medical Association and the American Institute of Aeronautics and Astronautics. His publications include articles evaluating life support systems for manned spacecraft.*

Judith Fey Thomas *is assistant director of the Phytotron and an assistant professor of soil science at North Carolina State University. She has been involved in experimental work and has published several articles emphasizing the effects of controlled environment on harvestable yields of various crops, in particular, the effects of daylength, temperature, nutrition and CO_2 levels on soybeans, tobacco and corn. In related work, she has examined the effect of mother plant environment on growth and development of the progeny.*

John A. Wilhelm *is a senior research associate in psychology at the University of Texas (Austin) and is a specialist in social psychology. He has been a project manager for various psychological research projects since 1970, and his principal interests are management of field research and management of data bases in the field of social psychology.*

Preface

The advent of the U.S. Space Shuttle and the increasing frequency of Soviet long duration space flights indicate that we are moving into a new stage of space activities. The question of basic human survival in space which was a substantive issue twenty years ago is now solved. Humans can survive and moreover work in space and this presents new opportunities and challenges.

The initial fifteen years of the space era was marked by the actual presence in space of a relatively few, physically very fit males functioning for short periods of time with a high degree of publicity and other forms of social reinforcement. Their activities were carefully planned beforehand and tightly supervised in orbit. In the Mercury, Gemini, and Apollo flights, spontaneity was all but non-existent. In Skylab and Salyut some spontaneity did exist but was severely circumscribed.

With the Space Shuttle and the long duration Soviet flights the texture of human activity will begin to change. There will be a greater number of people going into space -- estimates reaching into the hundreds before the end of the 1980s. Women and individuals from all races will be going. Until now, there was press attention to each and every space bound individual but this can be expected to fall off as well as the other types of social reinforcement that accompanied early space flight. Space will gradually become less exotic and more kindred to our ordinary endeavors.

New categories of work abilities will be required. The test pilots and scientists who have gone into space so far will be joined by lab technicians, engineers and eventually construction workers. Service in space will become industrial in addition to exploratory and the personality traits required there are expected to change. Emphasis will move gradually from the ability to handle high intensity, short duration challenges in a novel environment to long term execution of often repetitive duties that are without inspiration. An industrial job is still basically a job no matter if it is in Cleveland, the Alaskan North Slopes, or on a space platform.

The complete transition of space from the exotic to the prosaic will take some time but it is important to recognize now that we are starting down that road. This requires us to constantly update our conception of service in space and the kind of human qualities that are needed. Attention should be given to continually reshaping the selection and training procedures to the changing requirements of space work. Psycho/physiological knowledge can play an important part in selection. The qualities of dependability and social tolerance are as needed as a specific job skill. Training in interpersonal communication skills and in the art of leading, following and facilitating compromise in groups might be included along with the standard preparation for going into space.

The stay times in orbit may vary according to specific job functions, personality type, level of education, prior psychosocial history and family relationships. Beside baseline physical requirements such as radiation shielding and atmospheric composition, orbiting facilities may be designed to include psychological requirements. A person's need for privacy, to periodically change his or her environment or simply to be able to look out at the external world via windows may well need to be put into design considerations.

In the more immediate future special attention must be given to a person's natural posture in zero gravity, which is different from a gravity environment, and the increasing need to accommodate people of varying size. The early astronauts were selected to fit a fairly narrow range of height specifications, approximately 5'8" to adjust to the equipment. With greater numbers going into space being drawn from a variety of professions and nations the emphasis is now on designing equipment to accommodate a much broader range of height and other aspects of human physiognomy.

Orbiting facilities will be "total institutions" with all work and leisure taking place in one location. The on-board procedures will, by necessity, involve the governance, legal, mental health, social/cultural, financial, and communications aspects of the space workers' lives. Proper attention to the procedures can have a significant impact on the productivity and stay times of space workers. Poorly designed procedures will reduce productivity and stay times with potentially costly implications, given the high expense of moving people in and out of orbit. Well-designed procedures can significantly enhance the effectiveness of the overall operation.

The expanded size of space operations and the high cost of transporting material from earth require a fresh look at the feeding of space workers. What are the full possibilities for recycling waste into food and for that matter actually growing food in orbit? As futuristic as food growing in space may sound it must be studied if every avenue for efficiency and cost effectiveness in space operations is to be pursued.

Finally, what are the reasons for adding an industrial dimension to space? Are there economics involved that make the endeavor worthwhile? What goods and services can be obtained that make the economic development of space competitive with earth-based opportunities? These questions must be dealt with if the public investment necessary for space research and development is to be marshalled.

The editors of this volume have sought to stimulate reflection and research on the human factors involved in this intriguing epoch in space development through publication of papers by experts in the field. The papers are concerned with four basic areas: (1) the psychological aspects of working in space; (2) the design of orbiting facilities; (3) the production of food in space; and (4) the economics of industrial activity in space.

Psychological Aspects. Robert Helmreich, John A. Wilhelm and Thomas Runge provide an overview, discussing among other things future changes in crew composition and in the personal costs and rewards of spaceflight. The issues of crowding, privacy, leisure, authority structures, and male/female crews at a space facility are reviewed along with questions related to the selection of crews. Kirmach Natani, utilizing data derived from Air Force, space mission and Antarctic experience, examines "adaptive competence," that is, the ability of individuals to cope with immediate changes in the environment and to adjust to long-term changes while maintaining effective performance and continuing psychological growth. Adaptive competence is a vital element for successful service in space. Natani also discusses the utilization of neuropsychological techniques to assess adaptive competence in selecting people to go into space. Joan E. Sieber reviews the issue of privacy, outlining its various dimensions (self-ego, environmental, interpersonal, and control/choice) and its effect on personal stress and performance. She then discusses the difference between the individual crew member's need for privacy and the space facility administration's need to monitor crew health and performance.

Design. Richard Kline examines habitat design in view of possible missions during the 1980s and 1990s, which range from small work stations in low earth orbit that are occasionally manned to large permanently manned facilities at geosynchronous orbit. Allen J. Louviere and John T. Jackson look at the challenges to equipment design presented by anthropometric changes caused by weightlessness which include height growth, weight loss, and posture change. Equipment must be redesigned to accommodate these changes as well as the wider variety in size differences that the inclusion of women and members of other races and nationalities will introduce into space activities.

The Production of Food in Space. Paul Rambaut looks at human nutritional requirements in space in terms of the basic components of foods (fats, carbohydrates, amino acids, vitamins, etc.) and relates them to food production techniques in space (e.g., photosynthetic organisms). Michael Modell and Jack M. Spurlock present an overview of the major factors in the design of a closed food chain for extended space missions and provide a rationale for evaluating them. Marcus Karel discusses the research that will be necessary to plan food technology for future space habitats, which include understanding the various dimensions of human nutrition in isolated environments, the design of food production and preparation processes and equipment, and the social and legal factors implicit in food technology. C. David Raper, Jr., Terry A. Pollock, and Judith Fey Thomas present an overview of the uses of "phytotrons" in assessing the environmental requirements for plants in space habitats. Phytotrons are laboratories designed for the study of plant responses to environment and are defined as a collection of controlled-environment cabinets, rooms and glasshouses organized so that many combinations of independently variable environmental factors can be studied simultaneously. John M. Phillips discusses the possible application of controlled-environment agriculture (CEA) to future missions. CEA is a food production technology currently under development that is characterized by enclosed facilities where crop production factors are easily subjected to human manipulation.

The Economics of Industrial Activity in Space. B. P. Miller reviews the nature of the goods and services that can be provided by earth-orbiting satellites and also reviews related economic and policy issues such as costs, benefits and markets, and the interaction between the public and private sectors in the development and ownership of the satellite systems.

It is the hope of the editors that this volume will serve to encourage further work on the human factors in space, work which will parallel and complement the technological development of space. The editors have sought to stimulate research and publications in this area and will continue to encourage such activities in this important and growing field.

T. Stephen Cheston, Ph.D.
Georgetown University
and
The Institute for the
Social Science Study of Space

David L. Winter, M.D.
National Aeronautics and
Space Administration

Human Factors of Outer Space Production

Robert L. Helmreich, John A. Wilhelm, Thomas E. Runge

1. Psychological Considerations in Future Space Missions

Introduction

Since the suborbital flight of Allan Shepard in 1961, American astronauts have logged 938 man-days in space. The Russian program has added another 1,136 man-days, including the longest mission. Psychological adjustment to the conditions of space has been generally excellent, with no major problems reported during missions. Experience thus far would suggest that psychological factors will not be a major concern on future flights. Our thesis, however, is that generalization from early spaceflights can be misleading and dangerous and that *more* problems in psychological adjustment can be anticipated in the future. Russian space scientists have apparently drawn a similar inference from their program. Alexei Yeliseyev, the Russian space flight director, has asserted that the only barriers to long-duration spaceflights are psychological *(1)*.

We will not review the substantial literature on man's reactions in space but will instead outline the empirical and theoretical issues that lead us to expect more difficulties in adjustment unless preventive strategies are adopted *(2)*. The issues we will consider include the changing scope and goals of future missions, differences in crew composition, and shifts in the costs and rewards for participants. More speculatively, factors in crew selection and composition, authority structure, length of mission and the optimum physical environment will also be discussed.

The authors' research and preparation of this report were supported by NASA Grant NSG 2065 (Robert L. Helmreich, Principal Investigator). The opinions expressed herein are those of the authors and do not reflect policies of the National Aeronautics and Space Administration.

Factors in Personnel Requirements

New Missions in Space. Pioneering American spaceflights were manned by a cadre of highly trained, professional Astronauts. In the initial phases (Mercury and Gemini), those selected for the program came from the ranks of military test pilots. Some liberalization of selection policies occurred during the Apollo program with the recruitment of Scientist-Astronauts, individuals holding advanced degrees in a substantive area relevant to space exploration. In reality, however, the focus of early American spaceflight was on the high technology involved in safely sending men into orbit or to the moon, with science *per se* playing a distinctly secondary role. Indeed, the degree of technological expertise and pilot proficiency that Scientist-Astronauts were required to achieve in training for their flights appears to have had deleterious effects on their scientific careers (3).

In contrast, the forthcoming Space Shuttle program represents a major change in policy and a relaxation in requirements for flight. The Shuttle concept provides for a highly trained flight crew to be responsible for the launch, operation and re-entry of the orbiter. Additional crewmembers *without* flight qualifications or experience can now be accommodated. Thus, in this approach, scientists and others with work to perform in space can go into orbit without extensive technical training in the flight aspects of spaceflight. The result should be a considerable improvement in the quality of scientific research conducted in space, since scientists and other technical personnel at the forefront of their disciplines will have the potential to work in space without sacrificing commitment to their respective disciplines.

The shuttle-orbiter concept calls for a large number of relatively short (one week) orbital flights by the end of the 1980s, approaching a rate of one launch per week.

If the scientific potential of the next generation of space missions is fulfilled, the result may be the development of space stations dedicated to applied technology. Two applications for space stations are zero-gravity manufacturing and the transmission of solar energy to earth-based receiving stations. Should such industrial applications prove feasible, the type of orbital facility required would almost certainly entail a larger crew in orbit for extended periods. Thus, in long range planning, consideration should be given both to relatively small crews in orbit for short periods and larger crews in orbit for lengthy working periods.

Changes in Crew Composition. Whatever the evolution of the space program, one definite change will be in the composition of crews. Women have already been selected for the spaceflight program and will certainly be routinely going into space during the 1980s. In addition, crews for such programs as Spacelab will be more heterogeneous in terms of vocational interests and nationality, through collaboration with the European Space Agency. As was just mentioned, on many flights there will also be two categories of crew--flight personnel and scientific personnel. Such culturally and vocationally mixed crews are more likely to have interpersonal conflicts than the more homogeneous crews of the past.

If a long-term orbital facility devoted to manufacturing or energy transmission is developed, crews will doubtless become not only larger but even more heterogeneous. Such a station would almost certainly need workers with less education for relatively routine and repetitive tasks, in addition to highly trained scientific and technical personnel. Such crews might differ in terms of nationality, race, sex, social class, education, and vocational and avocational interests. Each of these factors raises the possibility of social conflict.

In summary, we can reasonably anticipate major changes in crews, which suggests that generalizations from the experiences of small, homogeneous crews with an extended, intensive backgrounding in the procedures of NASA may be quite unwarranted.

Changes in the Costs and Rewards of Spaceflight. Another factor that might lead to greater difficulty in adjustment on future flights involves the psychological costs and rewards associated with living and working in space. Reactions to unusual and stressful environments have been analyzed in terms of a cost-reward matrix *(4)*. The essence of this analysis is that outcomes (in terms of both performance and adjustment) are determined by an individual's *perceptions* of the costs and rewards inherent in the environment. Rewards are defined as the sum of all aspects of the situation that the participant sees as personally satisfying or as leading to future benefits. Costs are defined as the unpleasant components of the environment. These include the frustrations, discomforts, dangers, and all other foreseeable negative concomitants of living and working in a given situation. When rewards are perceived as outweighing costs and the situation appears, on balance, preferable to other options available to the individual, performance and adjustment are likely to be good, *even where the objective costs are very high.* This was clearly the case in the early phases of the American and Russian space programs. The costs were certainly high in terms of lengthy and grueling training, danger, and discomfort in the space capsule. The rewards were also great. Early Astronauts were national heroes, recipients of the world's adulation. Along with the hero status, a number of American Astronauts enjoyed considerable financial rewards and vocational opportunities outside the space program.

It is our contention that as pioneering enterprises become more routine and mundane, rewards tend to be more unstable than costs and to diminish at a greater rate. In the case of spaceflight, while costs in terms of the dangers and discomfort of missions will doubtless be reduced, missions will still be objectively dangerous and, for the foreseeable future, habitats will lack many of the comforts of earth. Rewards, on the other hand, can be expected to diminish even more dramatically. The sense of personal accomplishment and the extent of benefits from society should be markedly less for the 10,000th person in space than for the 10th or the first. As the exhilaration and challenge of the novel environment wane, participants' attention may begin to focus on the negative aspects of the environment that were endured or ignored during pioneering phases. Some evidence for the drop in extrinsic rewards can be seen in the increasing public indifference to space ventures after the first lunar landing. This is certainly to be expected when spaceflights become a weekly occurrence or when missions involve only the resupply of permanent orbital facilities.

<u>Crowding and Privacy in the Space Habitat</u>. With the exception of Skylab, American and Russian spacecraft have had very limited living space, making the term "crowded" an appropriate descriptor. Although visionary advocates of space colonization talk in terms of enormous orbital stations, a realistic appraisal of the costs per pound of orbiting payloads suggests that future vehicles and stations will continue to be environments of high social density.

While former Astronauts have commented on the extreme crowding in Gemini and Apollo capsules, adjustment to these conditions was, as noted, very good. This appears to be another instance where generalization from early flights to future situations could be dangerously misleading. The cost-reward analysis suggests that, as the rewards from spaceflight diminish, participants may be much more susceptible to the stresses of negative environmental conditions such as crowding.

Recent theoretical formulations on crowding confirm this assessment and suggest as well possible strategies to reduce the negative impact of high social density. Most workers in the area of crowding make a distinction between density, a physical measure of the number of occupants per cubic unit of space, and crowding, the psychological perception of an excess of undesirable social contact. Sommer, Hall, and others have defined personal space as an "invisible bubble," immediately surrounding the body, intrusions into which lead to anxiety and discomfort--the perception of crowding (5,6). Personal space is thus seen as an extension of the self. The size of this personal space in any one social contact is a function of the type of interaction and the relationship between the people who are interacting. Stress arises when what a person perceives as the appropriate distance is violated. This stress can be reduced by one of two means: rearranging the interaction in order to reestablish comfortable spacing or, when this is not possible or successful, through various coping strategies. Investigators have proposed a variety of mechanisms for coping which have differing implications for design. We will consider two of these which have both theoretical rationale and empirical support.

Worchel and Teddlie, who have proposed a two-stage theory of crowding based on the notion of personal space, have reported the results of a study relating to coping strategies (7). Subjects who had their personal spaces invaded and who were not given the opportunity to rearrange the interaction distance reported less crowding and performed better on a group task when the walls of the experimental room were decorated with visually stimulating materials (paintings and posters) than when they were blank. Their two-stage theoretical model explains these results in this way: the violation of personal space leads to physiological arousal. From Schachter's theory that emotions consist of cognitive labels applied to undifferentiated arousal (8), they argue that if an individual then *attributes* his or her arousal to the intrusion of personal space, he or she will experience the negative sense of being crowded. If, on the other hand, the arousal is attributed to some other external cause, there will be less of a sense of crowding, at least temporarily. Thus the subjects exposed to salient visual cues attributed part of the crowding arousal to the stimulating environment, not to social density.

Other examples of this misattribution process suggest an explanation of how positive emotions result from crowding. Individuals can be in a situation with very high social density, such as the crowd at an athletic event, without experiencing a negative sense of crowding. Presumably this is because persons in such settings attribute part of their arousal to the excitement of the event rather than to the violation of personal space. It may be for this reason that a sporting event is more exciting when viewed "live" than when seen on television. Similarly, early Astronauts may have made other attributions for arousal even in the close confines of the capsule. Astronauts undoubtedly experienced arousal from at least three major sources: fear, excitement arising out of the challenges of the mission, and extreme impingement on personal space. The most adaptive attribution for this arousal would, of course, be to the challenges of the mission. When missions become more routine and prolonged, attribution to *negative* sources of arousal such as crowding may become more frequent. The implication of Worchel and Teddlie's theory that may assist the designer is the derivation suggesting that, if an environment is complex and visually stimulating, inhabitants should tend to attribute part of their arousal *positively* to the physical environment rather than negatively to crowding. If dense situations are a reality, methods can be found to accentuate positively the coping process. For example, the designer can plan environments which are not only visually stimulating but also changeable. Creative use of lighting could be highly beneficial in providing novelty and making the environment more positive.

A related psychological variable is privacy, defined as a dynamic process that individuals or groups use to increase, decrease, or stabilize the amount of social interaction. According to Altman, privacy regulation is a function of the discrepancy between *achieved privacy* and *desired privacy (9)*. If achieved privacy is higher than desired privacy, loneliness or boredom will be the result. If on the other hand, achieved privacy is lower than desired privacy, the perception of crowding is the most likely outcome. Both states, loneliness as well as crowding, lead the individual or group to use certain strategies such as decreasing or increasing verbal or non-verbal communication in order to achieve the desired degree of privacy. These strategies involve physical and/or psychological energy expenditures which the person must maintain until the desired and achieved levels of privacy coincide. This effort is likely to be harmful to the individual as well as detrimental to task performance if it is not successful in a relatively short time. It therefore seems advisable to designate areas for either solitary or communal use, thus eliminating the necessity for stressful coping strategies. This plan will enable the crowded individual to withdraw, while still allowing other crewmembers (or the same individual at another time) to gather socially. Theoretically, highly flexible environments could help accommodate the needs for both privacy and sociality. Environments should be changeable and should have doors, signs, and other markers that either prohibit or encourage interpersonal exchange. Rather than having fixed, designer-assigned functions (for work, recreation, eating, sleeping, etc.) rooms should have their uses determined cooperatively by the inhabitants as far as is technically feasible. Screens, panels, and moveable furniture, for example, would provide considerable flexibility of space usage.

In making the environment changeable, provision should be included for modifying the physical arrangement of personal quarters (a particularly crucial point in long-duration missions with mixed crews). Over extended periods, strong and valid needs for privacy and intimacy are inevitable. Permitting individuals to have, for example, living and sleeping quarters that could easily be converted from single to double occupancy would not only provide variation but would also permit both solitude and more intense personal relationships (a point to which we will return later).

The usage of personal space differs widely between nationalities. For example, Latin Americans and North Americans define the boundaries of their personal zones at different distances, which sometimes leads to confusion and misunderstanding in social interactions between members of the two groups. Women tend to have smaller and more permeable personal spaces and to adjust better to close social interactions. Therefore, work space and living arrangements should be flexible to allow for differences in the use of personal space.

It is particularly desirable to provide the maximum possible visual access to the outside environment. Both Astronauts *(10)* and Aquanauts *(11)* have reported that the ability to observe the changing scene outside provided an extremely positive aspect of life in isolation and that the fascination that the visual environment held for them never waned. Observation of the outside environment should help crews redirect their attention from the more mundane world inside the space habitat.

Furthermore, under zero-gravity conditions crewmembers of Skylab reported that it was very difficult to adjust to an environment which lacked the vertical dimension; i.e., up and down were not defined. Technical mistakes are likely to occur if an Astronaut is not sure from which side he approaches a switch or a panel, and mistakenly pushes the wrong button. Appropriate designing can eliminate this problem by creating an artificial vertical dimension using graphical means such as arrows and lines to denote a "floor," lights suspended from a "ceiling," etc.

It may be necessary to teach future space crews how to change their environment. As Sommer has noted, most people do not interact actively with their physical environment *(12)*. Rather, they accept the status quo and react to a given surrounding. Spacecraft inhabitants should be made aware of possibilities for rearranging their physical environment to serve their needs and desires. They must understand that, in doing so, they are engaging in a reciprocal process: the crew influences the environment, which in turn affects their responses.

The designers of future space vehicles may have a more important task than did their predecessors. Not only must such facilities be safe and efficient, they must also satisfy the needs of more diverse, critical, and potentially disaffected inhabitants. It is probably safe to say that engineering considerations have been the primary determinants of spacecraft design in the past. If, however, psychological factors do play a greater role in future adaptation in space, then issues of *psychological* habitability must come to play a more central role in design.

<u>Leisure Aboard Spacecraft</u>. It is our belief that with increasing flight duration, the satisfaction of leisure needs will become more critical. Most missions to date have been relatively short and all have had extremely high workloads. The Astronauts of Skylab, particularly the third crew, had such a heavy workload that virtually no spare time was left over. It is noteworthy that not even one Astronaut finished more than one book during the entire 84-day mission *(13)*. Future space crews, with increased division of labor and more mundane tasks to perform, are likely to live according to schedules similar to those on ships, with an 8- to 10-hour working day. Under these circumstances, 6 to 8 hours per day remain (disregarding sleep) for meals, personal maintenance, and leisure. The scope of leisure activities possible on board a spacecraft will be restricted and will exclude many sports and outdoor activities. Substitutes will have to be found which are

satisfactory for most participants. Under zero-gravity conditions, large areas aboard the spacecraft allowed the Skylab crews to bounce off the walls, perform somersaults, etc. Cooper reports that this was a major source of enjoyment during the mission. It might be advisable to provide areas that permit this type of exercise.

Crews mixed on the basis of sex, nationality, and vocation will have very different leisure interests. A recent study by Runge and Helmreich indicates some sex differences in regard to leisure activities *(14)*. Female respondents reported a greater tendency to engage in expressive leisure activities, such as singing and dancing. Male respondents, on the other hand, reported more active leisure pursuits, such as sports. As space will be at a premium, it will be the task of mission planners to accommodate a range of leisure activities and facilities. It would appear highly advisable to routinely explore the leisure preferences of prospective crewmembers in the selection process. The design of leisure facilities could then be considered accordingly during mission planning.

Psychological Selection of Crews. A number of factors entered into the selection of early Astronauts. These included extensive flight experience (the original Mercury Astronauts were all test pilots) or scientific training (in the case of the later Scientist-Astronauts), physical condition, and psychological adjustment. It is probably fair to say that the primary psychological and psychiatric emphasis was on screening out psychopathology. Indeed, the early phases of the American space program have been aptly described as a space race with the Russians to obtain primacy in manned flight. The major requirement for Astronauts was the successful *testing* and operation of new spacecraft. Little stress appears to have been placed on arriving at the best psychological composition of specific crews drawn from the Astronaut pool.

Clearly, from the results of early missions--including reactions to the major emergency aboard Apollo 13--these selection procedures worked quite satisfactorily, although there is some evidence of postflight psychological difficulties *(15, 16)*. Because of the great challenges of the pioneering missions and the intense desire to succeed in opening a new frontier, Astronauts seem to have accommodated to one another rather well, and to have subordinated personal differences to the superordinate goals of the mission. However, in longer and more routine missions with more heterogeneous crews, individual idiosyncracies and group dynamics can be expected to assume a much larger role in determining the emotional climate during a flight.

If our analysis of the psychological costs and rewards of future space missions has even partial validity, *greater* emphasis on psychological selection and on crew composition is needed. In addition to screening against potential or actual psychological problems, selection should be for *positive* psychological characteristics. This, of course, presupposes that some consensus can be reached on optimal constellations of attributes.

Our research over the past few years has focused on the specification of positive dimensions of personality related to attainment and adjustment. In this endeavor we have concentrated on the study of masculine, instrumental traits and feminine, expressive traits, and on achievement motivation.

Masculinity and femininity were long considered to be the two end points of a psychological continuum, with all individuals falling somewhere along this continuum. This formulation bore the additional assumption that the possession of one attribute (i.e., masculinity) necessarily implies the lack of the other. A number of investigators have questioned the validity of this historic assumption and considerable research has supported a redefinition of the psychological natures of men and women *(17)*.

Psychological masculinity has recently come to be defined as a constellation of attributes denoting an instrumental, goal-seeking orientation, while femininity has been defined as a set of characteristics reflecting psychological expressivity and sensitivity to the feelings and needs of others *(18)*. The labels masculinity and femininity are justified because each cluster of attributes discriminates significantly between the sexes in normal populations, males scoring higher on instrumentality and females higher on expressivity. When these dimensions are measured independently, the attribute clusters prove to be essentially uncorrelated in each sex. That is, individuals may score high or low on each and a significant proportion of each sex scores high in *both* masculinity and femininity. The modal male, however, scores high in masculinity and relatively lower in femininity, while the modal female shows the opposite pattern.

The central empirical finding from this line of research is that those of both sexes who score high in masculinity and femininity (usually labeled as psychologically androgynous) appear to be advantaged in a number of important ways *(19, 20)*. In a variety of heterogeneous populations, we have found that androgynous males and females have more positive self-concepts and appear to be more interpersonally effective. Importantly, the distribution of masculinity and femininity in each sex appears to be quite impervious to social class effects and quite stable across the lifespan *(21, 22)*.

An important distinction must be made regarding these trait dimensions. The terms masculinity and femininity as used in this context refer to the limited set of attributes we have defined as instrumentality and expressivity. The fact that an individual scores high on masculinity and femininity does not lead to inferences about his or her preference for particular sex-linked roles or behaviors or sexual orientation. In fact, the relationships among psychological masculinity and femininity, role preferences, and sexual orientation are weak and complex *(23)*. As an example, in women the relationship between psychological masculinity and attainment in demanding, male-dominated professions is quite asymmetrical. A high proportion of successful women scientists score as being highly masculine on our measure of psychological masculinity and femininity, the Personal Attributes Questionnaire (PAQ). However, only a small percentage of highly masculine women enter "masculine" professions, the majority choosing instead to enact their instrumentality through successful fulfillment of more traditionally "feminine" roles. Thus, our narrower conception is that masculinity reflects the *capacity* for strongly goal-oriented, instrumental behavior, and femininity the capacity for sensitivity and emotional expressivity in interpersonal relationships. In this conceptualization, the psychologically androgynous person is *capable* of both effective goal-oriented behavior and effective interpersonal relationships, with the enactment of these attributes being determined to a considerable extent by situational and role demands.

Considering the future environment in space, these attributes would appear highly desirable for crewmembers. This is not to discount other dimensions on which personalities can be classified, but rather to note that a strong instrumentality combined with interpersonal sensitivity should be associated with task fulfillment and a successful social adjustment. Some data are beginning to emerge from laboratory research which support at least the latter notion through demonstration that androgynous individuals establish more rewarding social interactions *(24)*.

The second area we have examined encompasses the domain of achievement motivation. It has long been recognized that raw ability and intelligence, while predictive of attainment, leave much of the variability in performance unaccounted for and that much of the residual variance reflects individual motivation. However, attempts to operationalize and quantify the aspects of achievement motivation have not been notably successful (25). Most empirical and theoretical approaches to the assessment of the motive to achieve have conceptualized it as a unidimensional construct present to a greater or lesser degree in every individual, and have considered achievement motivation as the resultant of the algebraic combination of a unitary motive to succeed and a motive to avoid failure (26). More recently, however, the unidimensional nature of the motive to succeed has come under question (27). Our approach to this problem area has been to consider achievement motivation as a multidimensional construct and to attempt to measure the constituent parts reliably in each sex.

In research with respondents from a number of age and occupational groups, we have isolated three major motivational factors in each sex with a new, objective instrument: the Work and Family Orientation Questionnaire (WFOQ, 28). These factors and their associated scales have been designated Work Orientation, Mastery needs, and Competitiveness. The first, Work Orientation, contains items dealing with a positive orientation toward work as a rewarding endeavor. Mastery represents a desire to solve difficult problems and to better previous performance; conceptually, it represents a kind of *intrapersonal* competitiveness. The third factor, Competitiveness, defines a concern with bettering *others'* performance and winning in interpersonal situations. Not surprisingly, masculinity as defined and measured above is consistently and positively related to the components of achievement motivation (29). However, direct assessment of motives to achieve seems to result in better prediction of attainment than the use of masculinity scores. Conceptually, the achievement motives can be seen as more focused representations of instrumentality.

We have recently completed a series of studies relating measured achievement needs to criteria of attainment in several groups. Consistent findings showing an interactive relationship between the three factors and objective measures of performance were found in each case (30). In the first study, we determined that the highest college grades were found among those high in Work and Mastery needs but relatively *low* in interpersonal Competitiveness. More interestingly, the same relationship held in a sample of businessmen, using annual income as the criterion measure. Finally, in two samples of Ph.D.-holding scientists, the pattern of a positive correlation between Work and Mastery and attainment and a negative correlation between Competitiveness and the criterion was again replicated *(31, 32)*. The measure of achievement for scientists was the number of citations of their published research by other scientists, as quantified in the *Science Citation Index*. Predictably, those in each group low on all three dimensions showed the least achievement.

The finding that Competitiveness relates negatively to performance in a variety of groups is rather counterintuitive. Folk wisdom has keyed on competitiveness as the *sine qua non* of success in American society, and indeed competitiveness has been scored positively in unidimensional measures of achievement motivation. (This is probably one reason the unitary achievement motivation construct has not predicted well in the past.) It is our belief that concentration on the interpersonal aspects of competitiveness--on "winning"--can interfere with performance in otherwise highly motivated individuals by making them overly concerned with the possibility of failure. That is, they may tend to evaluate themselves primarily in terms of performance relative to others rather than in terms of their own objective performance. Perhaps rather than accepting the possibility of a position of secondary eminence in the eyes of others, they may inhibit the production of innovative work or may channel their energies into other competitive activities.

Another possible cause of the negative influence of competitiveness could be the disruption of interpersonal relations. High competitiveness may well lead to interpersonal stress and hostility, outcomes which could adversely affect collaborative work or the total work and social climate of an environment.

Anecdotal reports by Astronauts and journalists suggest that the early American Astronauts were highly competitive *(33, 34)*. Walter Cunningham states, "In whatever we did there existed that one universal thread: competition.... If we had to take a urine test, we competed to see who could fill the biggest bottle. (Once, when a nurse asked Schirra to leave a urine sample, he obtained a five gallon jug, filled it with water, poured in a bottle of iodine, and left it on her desk.)" It is likely that during missions this competitiveness was overtly suppressed in the interests of the program's goals. In the maturity of the space program, however, the deleterious effects of competitiveness are more likely to be experienced.

Because of the need for effective work performance *and* a satisfactory social environment, the high Work/high Mastery/low Competitiveness motivational constellation would seem to be optimal for space crews. Related to the earlier discussion of masculinity and femininity, our data suggest that androgynous individuals are more likely to manifest this pattern of motives, while high masculine-low feminine individuals of *both* sexes are more apt to be relatively high on all three dimensions.

Whatever the techniques of psychological selection ultimately employed in constituting future crews, the factors discussed here would seem to be relevant to both performance and adjustment in isolated environments.

Authority Structure. American space missions to date have had a vertical, military command structure. Overall control of flight plans and activities has rested initially with Launch Control at Cape Kennedy and, in flight, with Mission Control in Houston. Aboard the spacecraft, the Command Pilot has responsibility for all procedural and crew behavior, and he himself is directly under command of Mission Control. This has proved to be a generally satisfactory arrangement for several reasons. First, the overwhelming majority of American Astronauts came from military flight programs and were both familiar and comfortable with this hierarchical chain of command. Second, all Astronauts have undergone lengthy and homogeneous training within NASA in this type of organizational structure. Third, missions have had relatively well-defined and nonconflicting goals. This system has not, however, been completely free of conflict. It is noteworthy, though, that the problems with authority which have been anecdotally reported [for example, by Cooper in his documentation of the Skylab missions *(35)*] have centered on disagreements between distant, ground-based commanders and on-board flight personnel. Similar conflicts between mission crews and earth-based control were found in our studies of Aquanaut performance during the Project SEALAB and Project TEKTITE saturation dives.

Despite the demonstrated utility of this form of organization, past success is no guarantee that it will prove as effective in more complex and longer future missions. Problems with a rigid command structure are more likely on missions with more heterogeneous and less highly trained personnel. Many scientists, for example, are used to functioning with considerable autonomy in their own work and many civilians may be quite unfamiliar with military discipline.

The forthcoming Spacelab missions slated for the Space Shuttle already reflect a more complex organizational structure. On these missions, an Astronaut trained and under control of NASA's Astronaut Office will be in overall command and responsible for the integrity of the mission. However, the scientific program, the primary goal of such missions, will be controlled by a Mission Specialist to whom Payload Specialists will report. Futhermore, the specific scientific programs to be conducted will normally have ground-based Principal Investigators from a variety of industrial, academic, and government organizations. These Principal Investigators will also have the expectation of exerting some control over the execution of their experiments. Clear opportunities for conflicts of interest and difficulties in communication exist. Indeed, even in a ground-based test of the Spacelab concept involving a simulated seven-day mission, some difficulties were encountered (36). Clearly, considerable attention and sensitivity to conflicting goals will be required to minimize problems even on relatively short Shuttle missions.

Obviously, a strong centralized command is necessary to ensure the safety of missions and to respond to emergency situations, and this must be clearly understood and accepted by all flight personnel. Mission Control must be sensitive to the psychological issues involved in the remote exercise of authority. On longer missions involving more nearly self-sufficient vehicles, vesting a higher degree of autonomy in the space crew would appear desirable from the standpoint of both morale and productivity.

With larger crews on long-duration missions, it may prove advantageous to have several command structures: a strong leadership for mission control and flight safety decisions and a more democratic structure for everyday life and leisure. Control of scientific or industrial procedures which do not threaten mission integrity will probably be best vested in technical specialists, as in the Spacelab concept.

Accumulated data do not provide clear solutions as to optimal structures for the more varied missions to come. What is certainly required from NASA is close monitoring of the functioning of the authority structure in terms of both productivity and crew satisfaction, a willingness to experiment with different organizational structures, and the authority to immediately institute remedial changes if problems are seen developing either on board or between ground and space.

Sex. For a number of reasons, none of which are related to their ability to perform or adjust in the environment of space, women have been excluded from the American space program. (The Soviet Union sent a woman into space in 1963; in the U.S., during Project TEKTITE 2, a team of female Aquanauts performed at equal or superior levels to teams of male peers.) The exclusion of women has ended, as NASA now has women Astronauts in training. The central question is now the impact of sexually mixed crews on performance and adjustment in space.

Data suggest that in extremely stressful, isolated environments, sex is simply not a relevant dimension *(37)*. Other data, admittedly spotty and anecdotal, suggest that in highly motivated, work-oriented groups, sexually mixed crews perform admirably. There is therefore no reason to believe that short-duration, heavily work-oriented missions such as those envisioned in the Shuttle program will be at all influenced by the sexual composition of their crews. A possible but important exception to this would be the presence in crews of individuals with strong prejudices about the abilities and/or appropriate roles of the sexes. In fact, the openness of crewmembers to working with the opposite sex may well be an important consideration for selection.

Again, lack of sexual problems postulated for early, brief missions cannot be expected to generalize to missions with relatively large crews spending long periods in rather routine activities, with considerable blocks of free time. Under the latter conditions, intense sexual attractions appear inevitable. (Indeed, there are strong sexual pressures in groups of same-sex individuals similarly isolated for extended periods in monotonous environments, such as in prisons and boarding schools.)

Practically and theoretically, the best official policy for such eventualities would seem to be one of benign neglect. Legislation of personal morality in the confines of space would appear to be both futile and an unwarranted invasion of privacy. Selection of couples for long-duration missions would certainly *not* solve the problem of intense relationships, as demographic data demonstrate the fragility of dyadic relationships even without the constraints of danger, restricted mobility, and high social density. The best course of action in the cases of sexual bonding would seem to be to provide facilities for private relationships and to make such facilities easily modifiable so that changes in personal associations can be accomplished with minimal perturbation of the larger social network.

Assuming shifting sexual allegiances may appear to be an unduly cynical forecast of the social dynamics of mixed-sex crews on long-duration missions. Rather, it is meant to suggest the provisions that should be made to minimize the occurrence of intensely painful interpersonal conflicts in an inescapable environment.

Overall, we would suggest that in heterogeneous groups spending extended periods in a restricted environment, sexual and other interpersonal conflicts must be seen as necessary consequences of the situation, however successfully the crews are selected psychologically. The requirement for the social system is that it must provide social support for individuals with transitory psychological and interpersonal difficulties and must also provide easily attainable modifications in living arrangements. In other words, the social structure should be able to accommodate changing personal needs and desires.

<u>Mission Duration</u>. The foregoing discussion has dealt primarily with the relatively short missions scheduled for the Shuttle during the next decade but has considered the possibility that particular programs may develop which would entail long-duration missions with larger crews and more differentiation in roles. We have not discussed maximum feasible and optimal lengths for missions. This is a case in which there are few empirical data from space, the longest Russian mission having been 139 days and the longest American mission 84. However, to speculate a bit, a six-month mission with strong social support seems completely feasible from a psychological point of view. Using a cautious approach, close monitoring and thorough debriefing of a six-month crew would seem advisable before moving to periods of a year or more. It should also be noted that under conditions of interpersonal conflict and low social support, even a two-week mission could become extremely stressful.

One lesson which can be extrapolated from military experience is that indeterminate mission length is inadvisable (38). The indefinite rotation used in World War II was changed to a fixed, one-year combat tour by the time of the Vietnam war (39). Whatever the mission duration selected, it should be clearly understood by all participants and their families.

Long-duration missions raise the additional problem of what to do in the event of a psychological breakdown during a mission. In the case of short flights with small crews, a mission abort may be the only feasible course of action. The issue is considerably more complex on long flights with relatively large crews. In this case, immediate return may not be possible, especially from a permanent space facility. Bringing back a full crew could also levy heavy psychic costs on those whose planned work is terminated early. Thus, in the event of a psychological casualty, difficult decisions will have to be made by the authorities. Such decisions would require extensive data on the nature and origins of the psychological problem and on the current dynamics of the crew.

If possible, the optimal solution would seem to be to maintain the integrity of the mission and to treat the problem in space. Data from military psychiatry suggest that the more quickly a casualty is treated and returned to duty, the better the recovery. Such a procedure would, of course, require detailed contingency planning and precise data on both the symptoms and the treatment.

The Need for Research. It is almost mandatory to end any review or empirical report with a call for further research. In the present case, however, this call has a special urgency. We simply lack an adequate data base to enable us to forecast the reactions of the sorts of crews on the kinds of missions that may come to dominate the space program. We also need to acquire extensive data on reactions to missions of longer duration.

As we have noted, for example, leisure needs and activities have played a minimal role in the work-intense space missions to date. They will doubtless assume greater importance on longer flights. There are, of course, extensive data relevant to each of the considerations we have raised in this paper. What is lacking is information on how these factors might *interact* in the possible configurations of crews on the types of missions which may develop. This does not denigrate the value of extant data sources. Indeed, such environments as the Antarctic, nuclear submarine patrols, undersea saturation dives, oceanographic research voyages, and a number of other settings systematically or ethnographically studied provide provocative information on group performance and adjustment. And, of course one continues to look for earthly analogs to spaceflight. We are, for example, embarking on an observational study of supertanker crews because they have many of the relevant characteristics we have discussed here-- heterogeneity, long periods in relative social isolation, and mundane work assignments interspersed with real danger. Although it is an appealing research setting, data from this environment can be only suggestive of reactions likely in space.

When the first mixed-crew Space Shuttle is launched, the first directly applicable psychological data will be generated. The early phases of this program should generate the most face valid psychological data on adjustment and performance on this type of mission. The opportunity to obtain extensive psychological data should not be missed. Given the requirement for precise multivariate data and the impossibility of simulating all possible social dynamics in the space environment, what is urgently needed is a dynamic, ongoing psychological research program that will generate immediately usable data on individual reactions and group dynamics. Such data could be used for interventions much as engineering and navigational data are collected and used for in-flight corrections. Although social psychologists are unused to working in this mode, there is no reason why they cannot do so. Also crucial will be the attitudes of mission authorities toward such data and their willingness to utilize them. It is encouraging to note that the Life Sciences Advisory Committee of the NASA Advisory Council in a recent report shows awareness of and concern with the majority of these issues *(40)*. In any event, future missions should prove to be far more complex and interesting from a psychological point of view than those that have laid the groundwork for our utilization of space.

References and Notes

1. "Two Soviet Cosmonauts Land After Record 139 days in Orbit," *The Washington Post*, November 3, 1978, p. A14.

2. For general reviews and data summaries regarding men in space, the reader is referred to:

    (a) T.M. Fraser, *The Intangibles of Habitability During Long Missions* (NASA CR-1084, National Aeronautics and Space Administration, June 1968).

    (b) Space Science Board, *Human Factors in Long-Duration Spaceflight* (National Academy of Science, Washington, D.C., 1972).

(c) R.S. Johnston, L.F. Dietlein, C.A. Berry (Eds.), *Biomedical Results of Apollo* (NASA SP-368, National Aeronautics and Space Administration, Washington, D.C., 1975).

(d) R.S. Johnston and L.F. Dietlein (Eds.), *Biomedical Results From Skylab* (NASA SP-377, National Aeronautics and Space Administration, Washington, D.C., 1977).

For Astronauts' reports and well-documented journalistic accounts of missions, see:

(e) E.E. Aldrin, *Return to Earth* (Random House, New York, 1973).

(f) H.S.F. Cooper, *Thirteen: The Flight That Failed* (Dial, New York, 1973).

(g) M. Collins, *Carrying the Fire* (Ballantine, New York, 1974).

(h) H.S.F. Cooper, *A House in Space* (Holt, Rinehart, & Winston, New York, 1976).

(i) W. Cunningham, *The All-American Boys* (Macmillan, New York, 1977).

3. B. O'Leary, *The Making of an Ex-Astronaut* (Houghton & Mifflin, Boston, 1970).

4. (a) R. Radloff and R. Helmreich, *Groups Under Stress: Psychological Research in SEALAB II* (Appleton-Century-Crofts, New York, 1968).

   (b) R. Helmreich, "Evaluation of Environments: Behavioral Observations in an Undersea Habitat," In *Designing for Human Behavior*, J. Lang, C. Burnette, W. Moleski, C. Vachon, Eds. (Dowden, Hutchison & Ross, Stroudsberg, Pa., 1974).

5. R. Sommer, *Personal Space* (Prentice-Hall, Englewood Cliffs, N.J., 1969).

6. (a) E.T. Hall, *The Silent Language* (Doubleday, New York, 1959).

   (b) E.T. Hall, *The Hidden Dimension* (Doubleday, New York, 1966).

7. S. Worchel and C. Teddlie, "The Experience of Crowding: A Two-Factor Theory," *J. Personality Soc. Psych. 34*, (1976), pp. 30-40.

8. S. Schachter, *Emotion, Obesity and Crime* (Academic Press, New York, 1971).

9. I. Altman, *The Environment and Social Behavior* (Brooks/Cole, Monterey, Ca., 1975).

10. H.S.F. Cooper, *A House in Space* (Holt, Rinehart, & Winston, New York, 1976).

11. R. Helmreich, The TEKTITE II Human Behavior Program. In *TEKTITE II: Scientists in the Sea*, J.W. Miller, J.G. VanDerWalker, R.A. Waller, Eds. (Government Printing Office, Washington, D.C., 1971), pp. VIII-62.

12. R. Sommer, *Design Awareness* (Holt, Rinehart & Winston, New York, 1972).

13. H.S.F. Cooper, *A House in Space* (Holt, Rinehart, & Winston, New York, 1976).

14. T.E. Runge and R.L. Helmreich, "Leisure as a Function of Achievement Motivation and Masculinity-Femininity," manuscript in preparation.

15. E.E. Aldrin, *op. cit.*

16. N. Mailer, *Of a Fire on the Moon* (Little, Brown & Co., Boston, 1970).

17. J.T. Spence and R.L. Helmreich, *Masculinity and Femininity: Their Psychological Dimensions, Correlates and Antecedents* (University of Texas Press, Austin, 1978).

18. J.T. Spence and R.L. Helmreich, *ibid.*

19. J.T. Spence and R.L. Helmreich, *ibid.*

20. J.T. Spence, R.L. Helmreich, C.K. Holahan, "Negative and Positive Components of Psychological Masculinity and Femininity and Their Relationships to Neurotic and Acting Out Behaviors," *J. Personality Soc. Psych.*, in press.

21. J.T. Spence, R.L. Helmreich, *op. cit.*

22. J.T. Spence and R.L. Helmreich, "Comparison of Masculine and Feminine Personality Attributes and Sex-Role Attitudes Across Age Groups," submitted for publication.

23. J.T. Spence and R.L. Helmreich, "Masculine Instrumentality and Feminine Expressiveness: Their Relationships With Sex-Role Attitudes and Behaviors," submitted for publication.

24. W. Ickes and R.D. Barnes, "Boys and Girls Together--and Alienated: On Enacting Stereotyped Sex Roles in Mixed-Sex Dyads," *J. of Personality and Social Psychology 36* (1978), p. 669.

25. B. Weiner, "Achievement Strivings," In *Dimensions of Personality*, H. London and J.E. Exner, Jr., Eds. (Wiley, New York, 1978).

26. J.W. Atkinson and J.E. Raynor, *Personality Motivation and Achievement* (Hemisphere, Washington, D.C., 1978).

27. D.N. Jackson, S.A. Ahmed, N.S. Heapy, "Is Achievement Motivation a Unitary Construct?" *J. Re. Personality 10* (1976), pp. 1-21.

28. R.L. Helmreich and J.T. Spence, "The Work and Family Orientation Questionnaire: An Objective Instrument to Assess Achievement Motivation and Attitudes Toward Family and Career," JSAS *Catalog of Selected Documents in Psychology 8* 35 (1978), MS #1677.

29. R.L. Helmreich and J.T. Spence, *ibid*.

30. R.L. Helmreich and J.T. Spence, *ibid*.

31. R.L. Helmreich, W.E. Beane, G.W. Lucker, J.T. Spence, "Achievement Motivation and Scientific Attainment," *Personality Soc. Psych. Bull. 4* (1978), pp. 222-226.

32. R.L. Helmreich, J.T. Spence, W.E. Beane, K.A. Mathews, "Making It in Academic Psychology: Demographic and Personality Correlates of Eminence," manuscript in preparation.

33. W. Cunningham, *op. cit.*

34. N. Mailer, *op. cit.*

35. H.S.F. Cooper, *A House in Space* (Holt, Rinehart, & Winston, New York, 1976).

36. R.L. Helmreich, J. Wilhelm, T.A. Tanner, J.E. Sieber, S. Burgenbauch, "A Critical Review of the Life Sciences Project Management at Ames Research Center for the Spacelab Mission Development Test III" (NASA Technical Paper #1364, National Aeronautics and Space Administration, January 1979).

37. P.G. Zimbardo, *The Cognitive Control of Motivation: The Consequences of Choice and Dissonance* (Scott, Foresman, Glenview, Ill. 1969).

38. S.A. Stouffer, A.A. Lumsdaine, M.H. Lumsdaine, R.M. Williams, Jr., M.B. Smith, I.L. Jones, S.A. Star, L.S. Cottrell, Jr., *The American Soldier: Combat and Its Aftermath*, vol. 2 (Princeton University Press, Princeton, N.J., 1949).

39. P.G. Bourne, *Men, Stress and Vietnam* (Little, Brown & Co., New York, 1970).

40. NASA, *Future Directions for the Life Sciences in NASA* (Life Sciences Advisory Committee of the NASA Advisory Council, Washington, D.C., November 1978).

_____ *Kirmach Natani*

2. Future Directions for Selecting Personnel

Summary

A brief review of some apparent similarities between life at an Antarctic station during the winter and long-term space flight is presented. Both environments are novel and unique, and involve potential danger from both individual accidents and the failure of life support systems. Once established, the daily routine in both may become monotonous. However, due to the nature of these environments individuals may be required to suddenly shift from a state of boredom into their highest level of physical and mental activity to cope with an emergency. A certain type of individual is required to successfully meet these requirements to remain alert to potential hazards over an extended period of time and to maintain psychological adaptability under the influence of environmental stressors. Life in a high-risk environment requires an individual with psychological strength, an ability to learn quickly under unexpected conditions, tolerance for loneliness and anxiety, and an excellently functioning central nervous system.

The type of individual required is defined as being adaptively competent. Adaptive competence is the ability to cope with immediate changes in the environment and to adjust to long-term changes while maintaining effective performance and continuing psychological growth. Competence is a relatively new concept which is defined in terms of personal characteristics such as cognitive style, knowledge,

---

The author was supported during the preparation of this paper by a National Research Council Resident Research Associateship in Neuropsychology at the USAF School of Aerospace Medicine, Clinical Sciences Division, Brooks AFB, Texas 78235. Bryce O. Hartman, Ph.D., Scientific Advisor.

skills, attitudes, etc., that facilitate achievements having adaptive value in interactions with the environment. Competence is a consequence of learning and appears to derive primarily from the socialization process that develops the social-self. The end product, for the psychologically adaptive individual, is a lifestyle that provides satisfactorily for physical, emotional, cognitive, and philosophical needs within a socially productive context. Personal developmental history, future-self attitudes, stress testing, and peer evaluations appear to be the best methods now available for assessing competence. Assessments based upon evaluations of neuropsychological functioning related to developmental history may provide more objective measures in the future.

In terms of future directions for assessment it is suggested that measurements of functional lateral differentiation in the nervous system may be useful in testing neural efficiency as well as in providing an index of adaptive ability. It should be emphasized that much of the information concerning functional lateralization in the nervous system and its potential for predicting behavior is still highly speculative. Reliable techniques for the assessment of this aspect of neural functioning that can be used in an applied setting with normal individuals remain to be developed. Nevertheless, lateralization appears to be an important area for future work because the evidence available suggests that this functional feature of the nervous system may be directly related to the individual's past experience. Biographical background or past experience histories have consistently demonstrated a substantial relationship to valid predictions of future performance (1, 2, 3, 4, 5, 6). But comprehensive histories containing significant events in an individual's life are difficult to obtain and may require multiple interviews over an extended period of time. This procedure is often not practical. If assessment of functional lateralization can provide an objective measure of the influence of the individual's developmental history on his or her brain, it will provide an extremely useful tool for the prediction of behavior.

However, it should be noted that at this time it is difficult to assume that enough is known about the nature of cerebral functional specialization to choose an hypothesized right or left hemisphere task that will have high validity for all individuals who might be tested. Tasks oriented toward the function of a given hemisphere may also obscure the presence of a strong hemispheric bias or dispositional lateralization gradient. Therefore, the primary assessment criterion for lateralization should be essentially

equivalent performance in both hemispheres on the same
type of task. It is expected that symmetrical performance
will indicate well-integrated intrahemispheric functional
structures. Measurement of bilateral symmetry in performance is a standard technique in neuropsychological
assessment *(7, 8, 9)*, which appears to be more important
at this stage of knowledge of the mechanisms of cerebral
functions and interhemispheric interactions. Too much
emphasis has been placed upon the uniqueness of the two
hemispheres of the human brain, while too little is known
concerning interactions between them. The quality of the
neural mechanisms and their level of efficiency
within each hemisphere and how these factors influence
behavior appear to be more important than the recently
demonstrated differences that are now so prominent in the
literature *(10)*. Future research in this area needs to be
addressed to evaluating the functional significance of
interhemispheric interactions.

Both adaptive and nonadaptive individuals are attracted
to high-risk environments. Developmental studies suggest
that both types of individuals may change. Therefore, a
reassessment program is essential. Again, the development
of techniques for evaluating neural functioning appears to
offer one solution for the problem of periodic reassessment.
However, psychometric measures for assessing crew compatibility factors are also essential. The presence of socially
nonadaptive crewmembers in a closed, isolated group can
significantly decrease the group's efficiency. Problems of
cultural drift away from and hostility toward the supporting
society may also be increased by the presence of socially
nonadaptive individuals. To limit these problems, crewmembers should be dedicated professionals, time in isolation
should be limited, and support personnel should be veterans
of the isolated experience. Antarctic veterans may provide
a useful pool of potential volunteers for initial development activities in space. The most efficient method for
handling tasks of selection, training, isolated duty, and
support may be to have a formal occupational organization
in which individuals are rotated through duty in these
various phases of the operation during an extended career
commitment. Additional research in all aspects of assessment is required to assure the selection of those
individuals who will be most likely to complete their
mission safely and in good health.

Introduction

Interest in human factors relevant to the development
of space is now entering its third decade. Twenty years of

research and applied activities have made extraterrestrial missions of short duration a practical accomplishment for a few highly selected personnel. The historical and practical aspects of human accomplishments in space flight have been effectively summarized in several recent publications *(11, 12)*. Collins *(13)* and Rosen *(14)* have also reported some of the subjective aspects of life in space. A summary of relevant past research, with recommendations for potential problems of long-duration space flight, was prepared by a study group selected by the Space Science Board of the National Research Council *(15)*. The primary human factors problems of the future involve the selection of scientists, technologists, and skilled workers who have not had the extensive experience in aviation characteristic of the majority of present astronauts. These problems will be compounded by the extended duration of future missions, increased crew size, and greater heterogeneity of crewmembers in past experience and personalities.

Selection procedures in the past have relied upon assessments using psychometric testing and psychiatric evaluation, as well as skill levels and general physical fitness criteria. All of these procedures have been useful in initial screening and in the selecting-out of applicants who do not meet established requirements. In aviation, psychometric tests have had their greatest usefulness in reducing the number of trainees required to produce a given number of qualified graduates *(5)*. The procedures now in use are quite good for determining which individuals *could* perform a given assignment. They are much less accurate in determining who *should* be chosen *(6)*. It is extremely important to choose the right person for a given mission. Due to problems in the selecting-in of candidates from a pool of apparently well-qualified applicants, the final word in most selection procedures is an administrative decision *(16, 17, 18)* and often depends, ultimately, on how many individuals may be needed at the time *(6)*.

Before behavioral scientists can have more influence in the final decision, they will need a better appreciation of those factors being selected-for. They will also have to be able to provide an increasing number of valid dimensions of *positive* attributes for consideration, instead of primarily contributing information concerning potential weaknesses in a given candidate. The author has some first-hand familiarity with assessment procedures used in the selection of individuals for Peace Corps service and Antarctic duty *(19)*. Antarctic duty appears to have many parallels with long-term assignments in space. Therefore, the Antarctic experience will be used as a background for discussing some of the apparent similarities between the two environments and to illustrate how Antarctic research has influenced the author's perception of future developments in personnel selection.

Antarctic Research: An Untapped Resource

Crew Selection Studies

Results of 15 years of Antarctic research [20] suggest that crew selection and composition across diverse occupations such as those required in the construction and maintenance of a large satellite station in space will be a difficult and challenging task for behavioral scientists. Sells and Gunderson [21] examined the social aspects of these problems in some detail by using the results of research conducted at isolated Antarctic stations to develop a frame of reference for evaluating parameters of the complex microsociety that will evolve during life in a space ship or satellite station. Their efforts represent an exception, for the Antarctic literature has generally not been used in evaluating aerospace problems, even though it is highly relevant and extensive [22, 23, 24, 20, 25, 26, 27]. Stuhlinger [28, 29] suggested that the same motives which initiated and sustained polar explorations also impel man into space. Smith and Jones [30] made some interesting comparisons of personal autonomy between a number of men selected for the Mercury Program and Antarctic scientists. This study apparently did not generate a continuing interest in this type of research, even though Hartman and Flinn [31] suggested that Antarctic isolation studies provided data relevant to crew selection procedures for space missions. More interest in the use of knowledge obtained from research in the Antarctic may result from the description, by Shurley, Natani, and Sengel [32], of potential psychosocial problems facing the first technology satellite crew. Their paper assumed no significant improvements from then-current techniques in assessing adaptive ability.

Summary volumes and translations of Russian Antarctic research have appeared relatively recently, a fact which may have contributed to a neglect of the results of this work. Many human studies conducted in Antarctica appear to be related primarily to the study of physiological adaptation. Thus, the behavioral scientist not sensitive to brain-behavior relationships or not interested in neurophysiological aspects of behavior might not be familiar with this literature. On the other hand, the Antarctic literature may have been discounted because many investigators, especially university-based behavioral scientists, believe that field research may be inherently inferior to laboratory research (33). However, the observations of Simons, Henderson, and Riehl (34) suggest that even though field studies may provide data that are highly anecdotal, the experiences offer valuable insights that cannot be obtained under laboratory conditions. Radloff (35) makes similar points, and indicates in addition that subjects often drop out of laboratory simulations due to boredom. Work in the Antarctic has confirmed the value of field research (36,37).

Antarctica as a Field Laboratory

The Antarctic work was conducted within the U. S. Antarctic biomedical research program developed under the leadership of Shurley (38,39,40,41). This work included an attempt to obtain psychometric measures of the influence of the multitude of stresses present in the Antarctic environment upon individual psychological functioning. Reports of the effects on the men in this and other situations of confinement typically yield clinical observations consistently suggesting a depressive syndrome, or "dysadaptation neurosis" (42), characterized by sleep disturbances, reappearance of symptoms of old chronic diseases, general lassitude, sluggishness and reduced working capacity coupled with irritability and apathy as well as occasional groundless fears. Soviet researchers believe these responses are primarily due to exhaustion of the central nervous system (CNS) coupled with cardiovascular insufficiency. The cardiovascular insufficiency appears to be related to the confinement and resulting hypokinesia (43) occurring in the Antarctic. Similar subjective symptoms have also been reported by subjects participating in sensory isolation studies, although long-term studies have not produced data to substantiate many of the feelings reported (44). It is interesting to note that a subjective symptom cluster of this type has also been associated with exposures to toxic agents (45) and with fatigue effects (46,47).

The resistance of the Antarctic syndrome to quantification and description by a number of commonly used psychometric techniques has intrigued us for some time. Administration of the Minnesota Multiphasic Personality Inventory (MMPI) on two occasions at the Amundsen-Scott South Pole Station during the Antarctic winter (48) confirmed the conclusions of Smith's (44) review. The MMPI profiles obtained were normal and showed no significant changes between the two administrations. Testing performed in a more severe environment at Plateau Station, which is at a higher altitude than the station at the South Pole, indicated some deleterious effects but the overall picture was one of significant improvement in psychological adjustment between the MMPI administrations (49)! In their appearance and behavior these men showed signs of ranging from, at best, clinical depression to, at worst, a state of profound chronic fatigue.

These unexpected MMPI findings occurred in the presence of significant changes in sleep patterns at Amundsen-Scott Station (50) and both objective as well as subjective observations of deleterious effects upon mood and performance. The mood and performance findings have only been published in summary form (48). The results obtained suggest that the MMPI does not provide a good measure of dynamic adaptive ability, although the positive indications of adjustment may be valid. This lack of sensitivity for the prediction of behavior may be true of other, or perhaps all, self-report psychometric instruments. The available data suggest that the men adjusted to this unusual environment, but the evidence of adjustment was not an indication that they would respond appropriately under the additional stress of an emergency. Circumstances provided an opportunity to observe behavior in emergencies on two occasions: during a fire and during forced evacuation of the station by the entire crew. During the fire, several men ignored safety procedures and put themselves at risk for serious injury from frostbite. The evacuation, which was planned in advance, proceeded smoothly but generated stress later in terms of how certain men evaluated their performance during preparation for the evacuation. These incidents demonstrated again that where time is a critical factor more errors in adaptive response will occur.

During the first five years of the Antarctic studies, the major effort of the research program was an analysis of the electroencephalographic (EEG) sleep data collected at six points in time during the subjects' participation in the study. Shurley and Pierce wisely selected sleep as the primary dependent variable because it provided an objective, quantifiable measure of all the forces acting upon the individual: psychological, social, and environmental. Recent publications of results of EEG frequency analysis (51,52) have been directed to a Soviet audience because, so far, the Russians have performed the most work in this area. This is consistent with the Russian treatment of problems of human spaceflight in the context of a comprehensive biomedical research program (53). The U. S. effort, on the other hand, has approached the task as one requiring an empirical development of human performance capabilities (54).

Lester, Burch and Dossett *(55)* reported that deep, or slow-wave, sleep is virtually eliminated during a state of subjective stress, such as the period before a critical examination for medical students. Usually, after an acute stress of this type, there is a rebound of slow-wave sleep on succeeding nights that may compensate for the loss while under stress. Hartman *(56)* and Harris, Pegram and Hartman *(57)* have found reductions in deep sleep during extended flying missions. They have interpreted this effect and the subsequent length of time taken for recovery as an index of depletion of physical reserves or as the physical cost of a stress-inducing activity. The Antarctic work suggests that a year or more is required before slow-wave sleep returns to baseline levels in the men who winter at the South Pole. Thus, the men at the Amundsen-Scott Station may be experiencing a continual depletion of reserves with an accumulating state of chronic fatigue. This view is consistent with the Soviet suggestion that CNS exhaustion occurs in the Antarctic. The similarity of the subjective complaints obtained under conditions of sensory isolation/deprivation, Antarctic wintering, toxicity, and fatigue suggests that all of these effects have a common neurophysiological basis, i.e., a lowered neural efficiency, which is subjectively experienced primarily as a disturbance in sleep quality with feelings of apathy, irritability, lassitude and negative mood. The work in the Antarctic suggests that as long as a significant level of stress exists, sleep patterns will remain disrupted and prevent recovery, leading to a continual depletion of physical reserves.

Hypoxic Stress and New Learning

Among other things, the Amundsen-Scott Station at the geographical south pole is a high-altitude observatory with a mean annual barometric pressure of 500 mm Hg (680 millibars), or an equivalent physiological altitude of approximately 3350 meters (11,000 feet). The attempt to understand the relationship between residence in this hypoxic environment, sleep, and behavior stimulated a review of studies of performance conducted at low ambient oxygen tension in both field and low-pressure chamber experiments. The conclusion produced by this unpublished review was that the failure to find significant impairment of psychomotor and/or cognitive performance at mild levels of hypoxic stress was due to a combination of insensitive measures, poor experimental design, and inappropriate data anlysis procedures. Several reports support this conclusion *(58,59, 60,61,62)*. With proper control of practice effects and individual differences in response to hypoxia, significant decrements in choice reaction time *(63)* and in learning *(59)* have been demonstrated at pressure levels equivalent to 5,000 feet. These findings suggest that one of the major reasons for lack of more substantive data documenting the effects of subtle stress effects has been the use of inappropriate measures and procedures *(64)*.

The hypoxia review also established an acquaintanceship with the work of Warren Teichner *(65)*. Teichner was interested in developing a systematic approach to the prediction of human performance that would include both task variables and environmental factors. Teichner *(45)* used hypoxia (hypoxemia induced by exposure to carbon monoxide) as an environmental stressor. Hartman *(66)* also used hypoxia (induced with an altered oxygen-nitrogen breathing mixture) in assessing stress tolerance. Teichner found that initial learning conditions and response variability were the most sensitive measures. Hartman *(47)* also noted the increase of response variability under conditions of fatigue. Both Teichner and Hartman point out that an analysis based upon average performance levels may tend to cancel out subtle stress effects which appear primarily as increases in the variability of response latency and occasional lapses in response production or blocking. Teichner *(67)* emphasized that it is necessary to measure controlling processes instead of controlled events. Self-reported responses to items on most psychometric instruments are highly controlled events. They provide little insight concerning the dynamics of controlling processes occurring within the individual. Teichner *(67)* and Teichner and Olson *(68)* suggested that multiple input-multiple response tasks would be particularly sensitive to changes in selective attention *(69)* related to physiological compensatory reactions or controlling events.

Task Sensitization with Lateralization Techniques

Through close association with Parsons *(70)*, the author developed an interest in lateralization in the nervous system and the possibility of increasing the sensitivity of neuropsychological assessment procedures by lateralizing the test stimuli. In 1974, work was initiated on the development of a lateralized task that would be sensitive to subtle deficits in neuropsychological functioning. Preliminary results with the technique developed suggest that it has potential for clinical neuropsychological assessment *(71)*. Further development of the procedure for increased ease of administration and enhancement of data reliability has been achieved *(72)*.

Increasing familiarization with the literature and concepts of lateralization, as well as practical experience working in this area, suggested that the attentional bandwidth phenomena postulated by Teichner could be interpreted in terms of cerebral lateralization *(73)*. Teichner *(68)* has indicated that the psychophysiological state of the organism can be assessed by its attentional disposition. An attentional state of narrow bandwidth is presumed to reflect an activated state while a wide bandwidth is associated with lower levels of activation; i.e., bandwidth is inversely related to activation level *(74,67)*. The lateralization concept assigns a serial information-processing role to the left hemisphere. By definition, serial processing is a narrowband process. In this same conceptual framework, the right hemisphere has been hypothesized to function in a parallel processing mode. Again, by definition, parallel processing is a wideband phenomenon. Thus, one would expect that the quality of information processing as defined in terms of functional lateralization theory would be correlated with an individual's activation level. There is some evidence to suggest that such a correlation does exist and that body position can modulate both activation level and the state of hemispheric dominance *(75)*.

Teichner further suggested that, with changes in activation level and/or changes in information processing demands, the individual would shift between a narrow and a wide bandwidth orientation depending upon specific task variables and environmental factors. He hypothesized that response blocking reflects reversals in the attentional process and that blocks occur during these shifts in attentional state. Changes in response latency, i.e., increases in variability, were also assumed to be associated with these fluctuations in attention. Teichner clearly did not tie the criterion state or rate of reversal directly to activation level. Instead, he proposed that shifts in bandwidth were related to critical levels of input/output rate ratios tied to the task demands placed upon the central nervous system (CNS). Teichner also suggested that conditions could occur in which reversals would not appear. In this case, information processing in either the wide or narrow band condition would be optimal at some level of input but become inefficient at other levels. Thus, he was able to restate the inverted U-shaped performance function, usually associated with activation level, in terms of a data-processing ratio. Conditions or individual propensities that favor attentional processes confined to one type of bandwidth may also be interpreted in terms of "hemisphericity," as defined by Bogen *(76)*. Levy & Trevarthen *(77)* have referred to this effect as "dispositional lateralization." The concept suggests that lateralized biases in information processing, or attentional orientation, may provide strong and relatively stable influences upon human behavior. Thus, the key to the description and analysis of the neural basis for psychological adaptability may reside in the differential specializations of the two hemispheres of the human brain.

Cerebral Lateralization and Adaptive Ability

Laterality studies currently in progress suggest that there is a base rate of alternation in the human brain between momentary states of wide and narrow band information processing, or fluctuation in control between the left and right hemispheres for attending to the environment. This alternation appears to be under the control of a subcortical or brainstem switching mechanism that is sensitive to levels of both activation and neural efficiency. The operational properties of this subcortical system may be determined primarily by the individual's genetic background and psychophysiological status. Changes in activation level or neuropsychological integrity (due to the effects of hypoxia, fatigue, toxic states, etc.) could affect the optimal switching rate by biasing the circuit to favor one hemisphere over the other and/or interrupting the flow of interhemispheric information transfer. Both effects would have a negative influence on information processing efficiency. Thus, as Teichner postulated, activation effects may exert their influence at a level that is not directly concerned with information processing but that can significantly influence the input/output rate ratio. Under conditions in which the individual is comfortable, or adapted, there would be minimal interaction between the switching system and the interhemispheric system (with information sharing occurring between the two hemispheres) and blocks would rarely be seen unless special procedures are used to detect them (78).

At the information processing level, the functional specialization of each hemisphere (developed in terms of wide and narrow bandwidth processes) may not determine which hemisphere will eventually assume dominance for a given task. Evidently the past experience of the individual results in the establishment of a metacontrol system which provides a dispositional lateralization gradient (79,77,80). Exploratory efforts in the assessment of such a disposition, or state of functional hemisphericity, suggest that there are at least three discrete patterns of cerebral organization in normal right-handed males. The most adaptive organization appears to be one that yields no significant differences between the two hemispheres in performance measures on the primary task used (72). This is evidently the case when both hemispheres are highly differentiated and the individual is capable of swift and flexible alternation between response sets requiring the specialized abilities of the right or left hemisphere. The other two organizations indicated suggest less differentiated patterns, with either some right hemisphere functions in the left side or some left hemisphere functions on the right side. A cerebral organization possessing less differentiation and more mixing of specializations in the two hemispheres would tend to generate interhemispheric conflicts under levels of high information load. This, in turn, would increase activation levels and could establish a conflict between attentional shift rates and the need to exercise behavioral programs localized in a given hemisphere as a result of training. The net result would appear as response blocking and, occasionally, as inappropriate behavior. In some extreme cases, behavior might deteriorate to the level seen during initial learning as one hemisphere attempts to cope with a task for which a response output program has become localized in the opposite side due to training and is unavailable due to a stress-induced dispositional lateralization.

In addition to the potential for interhemispheric conflict and deleterious additive effects due to an increase in activation, an individual possessing less differentiation and more mixing of specializations in the two hemispheres may be required to make some type of compensation for this lack of differentiation. The concept of compensation also suggests that compensating individuals will have to mobilize their neural resources to a greater extent and work harder than noncompensators to attain comparable levels of performance; i.e., they will be in a state of higher activation. Thus, the compensator may also be susceptible to more fatigue and stress than a noncompensator. Inappropriate strategies due to functional dispositions for a particular mode of information processing could interfere with adaptive flexibility and also impose greater cortical workloads. The behavior produced in performing a given task therefore appears to be the result of a complex set of interactions between factors related to the individual's past experience, cerebral differentiation, the properties of an hypothesized subcortical attentional switching mechanism, and activation level. Thus, when environmental supports for established cognitive systems are altered or fail to be useful in a novel situation, one would expect to see marked individual differences in response resulting from differences in these neural substrates of behavior.

The use of laterality factors in selection procedures may be extremely useful. Once reliable procedures have been established it should be possible, on the basis of cerebral lateralization, to select individuals who will be more resistant to the potential degrading effects of stress. It may also be possible to develop remedial measures for individuals identified as potentially at risk. However, much of what has been presented here is still highly speculative. The major work remains to be accomplished, and it will be challenging. Some time ago, Jasper *(81)* concluded that no single clear-cut measure of lateralization in the CNS would ever be found. Lateralization appears to be a dynamic and elusive property of the nervous system and quantification of this aspect of the human brain for practical applications will require the efforts of more than one investigator or research group. Research efforts on several fronts by many individuals are required. Fortunately, recent developments in psychometrics and neuropsychology, coupled with the application of computer technology in the behavioral sciences which has led to the establishment of neurometric procedures, suggest independent, converging lines of investigation of adaptive behavior that can be applied to the solution of this problem.

Psychological Adaptability and Competence

Adaptive competence may be defined as the ability to cope--i.e., to possess a characteristic tendency for making an appropriate response under difficult, emergency, or novel circumstances *(82)*. This is the short-term requirement, in the context of performance and reliability in a high-risk environment. In addition to the immediate ability to cope, adaptive behavior also involves effective adjustment to both natural and man-made aspects of the environment. In the long term, adaptation may require delayed action, temporary retreat from a problem, abandoning previous strategies or positions, acquiring new knowledge and developing a new method of approach *(82)*. Adaptive behavior is evaluated by determining the degree to which an individual functions and maintains him or herself independently while satisfactorily meeting culturally imposed demands of personal and social responsibility *(83)*. Effective adaptive ability can be conceptualized in terms of competence, a concept developed by White *(84,85,82)* that focuses on an individual's functional needs. White *(85)* defined competence as the cumulative result of an individual's history of interactions with the environment. Competence can be evaluated by assessing self-attitudes, world attitudes, coping/behavioral/ cognitive styles, future-self orientation and the capacity for making a large but disinterested personal investment in a challenging undertaking *(86,87,88,89,84)*. White *(82)* emphasizes that adaptive behavior involves simultaneous management of a number of factors. An individual must continue to obtain and integrate information from the environment, maintain internal conditions of psychophysiological stability, and remain sufficiently autonomous to be free to exercise flexibility of movement and action.

Competence is a relatively new concept. Unfortunately, for many it appears to be a vague term that can be used to denote a number of facets of human functioning such as motivation, self-perception, social skills, occupational skills, and legal status. Measures of adaptive ability in terms of competence for independent maintenance and behavior have been developed as a supplement to intelligence measures for the mentally retarded *(83)*. For some, this suggests that competence is best applied in the assessment of minimal skill levels and is not useful for evaluating high levels of creative adaptation. An alternative definition from that offered by White includes all of the above but still emphasizes adaptation. Sundberg, Snowden, and Reynolds *(90)* define competence as . . . "personal characteristics (knowledge, skills and attitudes) which lead to achievements having adaptive payoffs in significant environments." This definition emphasizes accomplishment, personal resources related to adaptive ability, and interactions between behavior and the environment. Past accomplishments can be evaluated and are predictive of future achievements. Some measure of potential behavior-environment interactions can be obtained with task performance in the laboratory, but individual sources of adaptive ability present the greatest problem for measurement and evaluation at this time. White *(85)* suggested that competence is largely a consequence of learning--i.e., it is developmental in its scope. Thus, the assessment of developmental factors in adult behavior suggests possibilities for predicting adaptive behavior. This approach was taken by Fine and Jennings *(3, 91)* and by Vaillant *(92)*. Fine and Jennings *(91)* summarized modern developmental theory in four basic concepts: (a) idiosyncratic individual styles of adaptation derive from a developmental sequence that is universal; (b) satisfaction of emotional needs is as important as life events in shaping development; (c) psychological growth depends on transactions with other people; and (d) there are both genetic and social factors involved in a specific developmental sequence. Vaillant has emphasized the social aspects of development and indicated that it is the quality of sustained relationships with significant people that develops psychological adaptability and shapes an individual's future. Fine and

Jennings concluded that knowledge of an individual's developmental history will provide one with the necessary information to predict behavior under specific conditions and to match individuals with jobs. Some evidence to support this conclusion is provided by Reinhardt *(93)*, who found that superior jet pilots in the Navy were first-born children with a history of unusually close father-son relationships. Vaillant *(92)* concluded that adaptive ability involves both biological and psychosocial factors. He suggested that the most important biological factor may be the state of the nervous system and the manner in which experience has influenced the development of the brain. This position is consistent with the author's view that it may eventually be possible to obtain an objective evaluation of the influence of developmental history upon an individual's adaptive ability by measuring neural functions.

Selecting for Competence

Given time and opportunity for personal contact, a good coach, flight instructor, or expedition leader can identify individuals who do not possess characteristics of psychological adaptability and competence. Unfortunately, psychologists and psychiatrists, when limited to a single interview, do not appear to be able to match the selection performance of these seasoned veterans *(6,17,18,89)*. Smith *(89)* found that psychiatric mental health ratings had no correlation with actual performance in a field setting. Perry *(17)* suggests that the average behavioral scientist has little appreciation of specific factors being selected-for. It is also important to note here that Ezekiel *(86)* found significant sociocultural differences in the predictive validity of future-self assessments taken in the form of fictional autobiographies written by the individuals being evaluated. He also obtained evidence to suggest that autobiographical style duplicated interview style. Thus, for some individuals an interview may be an ineffective method of obtaining information useful in predicting their future performance. This view is supported by Mechanic's *(94)* suggestion that the attitudes and perspectives necessary for psychological adaptability may have developed so slowly and subtly that a person may not know how he actually feels about certain issues or how he would react under given conditions. Interviews are also limited by the problem of socially desirable responses. Therefore, they may produce information relevant only to the interviewee's perception of the purpose of the interview. As Hills *(95)* has indicated, once typical college students have learned what is expected of them, they can handle an interview with composure.

To a large extent, personnel selection by an expedition leader or group of potential peers familiar with the demands of the environment was the method used to select men for early Antarctic expeditions *(96)*. In many instances this method is still used to recruit and select personnel for polar duty. Carpenter *(97)* reported a similar technique for recruiting Sealab personnel and suggested that elaborate selection criteria and testing procedures are not necessary for forming small crews. Astronaut selection from the ranks of experienced test pilots was based on the same assumption; i.e., that relevant past experience on the part of the recruiter and/or the recruit would be most predictive of success *(98)*. Unfortunately, this method of one-on-one selection becomes impractical when large numbers of individuals need to be selected-in rather than screened-out before participating in a costly training program. Under these conditions, psychometric methods have demonstrated validity only for screening out risky choices. The quantitative differentiation of adaptively competent individuals within the screened pool of recruits may not yet be within the capabilities of behavioral science. Flinn *(98)* suggested that accuracy in assessing adaptive competence may never achieve a level comparable to the accuracy of assessment now available for predicting aptitude for learning complex technical skills. One hopes that developments in computer-assisted assessment combining psychometric, neurometric, and neuropsychological techniques will provide a more practical means for the selection of competence in the future.

Observations from the Field

What type of individuals are attracted to a profession or sport that involves more risk than one usually exposes oneself to during the commuter's five o'clock *grand prix*? The author's experience outside the laboratory in the field, or the real world, suggests that two basically different types of people may be found in the high-risk environment. The first type of individuals appears to have a psychological need to put themselves in dangerous situations and simultaneously deny the realities of the risk. They may or may not be thrill- or sensation-seekers *(99)*. Their positive self-esteem appears to depend upon their ability to demonstrate mastery of dangerous situations. They may also be highly influenced by the "romantic" aspects of certain occupations or endeavors, particularly those perceived as demonstrating power and invulnerability. As Smith and Jones *(30)* found, maladaptive poor performers may be more concerned with their role image, i.e., what they appear to be, than with performance of the tasks required by that role. Evidently, this apparent drive for mastery or control is too often a need for control over themselves that leads to the neglect of their immediate tasks or the environment. Denial of risk is a constant feature of the potentially self-destructive behaviors this personality type engages in: behavior such as non-use or improper use of safety equipment, protective clothing, and survival gear; discounting potential deleterious effects of fatigue, alcohol, and medications upon performance; and neglect of established public health measures such as avoiding suspect foods, boiling drinking water, and taking prophylactic medications. Their motivations are counterphobic *(100)* and are not based on a realistic appraisal of the environmental risks and rewards.

The second type of individual, those with an adaptively competent personality, consists of action-oriented problem solvers who, having mastered themselves, are interested in mastering and controlling a specific segment of their environment. The problems of interest may be self-generated, highly technical, or related to basic science. The applied problem solver appears to have motivations similar to those of the basic scientist. Motivation is an important issue *(101)*. Adaptive persons will persist in the face of adversity, failure, and frustration because they have developed well-defined goals and realistic cost/reward expectations, while maladaptive persons may reevaluate their goals and decide their interests lie elsewhere when life becomes "uncomfortable." The individual with the competent personality type also appears to possess slightly obsessive-compulsive traits, when operating at risk, and tends to take nothing for granted. These individuals will make full use of any equipment or techniques available that will reduce the level of risk to which they are exposed. They much prefer to do things for themselves if the tasks are relevant to their safety and survival.

Smith *(89)* has found two distinct personality types among good performers in a field setting. Only one of these types would be classified here as being ideally adaptively competent. Because there is more than one subtype, Smith suggests that selection policies based on a stereotyped conception of the ideal adaptive individual could have poor predictive validity in terms of identifying all potential good performers. He identified one type of individual possessing role-appropriate skills and attitudes coupled with a pattern of self-confidence, commitment, energy, responsibility, autonomy, flexibility, and hopeful realism. Smith's other type of good performer was a somewhat rigid, low-commitment "9-to-5'er", who could become sufficiently involved in the work to produce a significant contribution.

Smith described individuals that would be classified as potentially maladaptive: hardworking but unconventional, impulsive, confused and chaotic, not sure of the future or themselves, and primarily interested in self-cultivation and personal improvement.

These differences in behavioral and motivational orientation have profound implications for effective performance in real-world situations such as those involved in expanding new frontiers under the sea, in the polar regions, and in space. Identification of individuals appropriately suited to these endeavors is extremely difficult because maladaptive characterological traits are often not obvious in what would be described as a normal, less stressful, environment. Some individuals may show undesirable or maladaptive behavior only under conditions of chronic stress, or the apparently exemplary person may be prone to cyclic or episodic periods of nonadaptive behavior. The most desirable individual may combine traits of both the ideal stereotype and the 9-to-5'er. Individuals possessing traits of the ideal stereotype may find greater rewards in being a pathfinder and may tend to become bored and less effective when the major problems challenging new endeavors have been satisfactorily resolved.

Continuing Reassessment

Nonadaptive behavior is too often seen in aviation. Unfortunately, this behavior is not limited to the amateur pilot. Pilot error, classified as a psychophysiological factor, is involved in 45 percent of all aircraft accidents (102,103,104). Accident reports suggest that even highly experienced commercial pilots may behave maladaptively under time stress (105,106) and in response to weather-related problems (107). The novice or characteristically nonadaptive individual may experience difficulty in coping with a novel environment. However, the problems of the professional aviator appear to be related to the loss of the ability to perceive novelty in what appears to be a familiar setting. As Livingston (53) has noted, complex activities have essentially unlimited potential for novelty while an individual's previous experience, for even the most seasoned veteran, is inevitably severely limited. Habituation to the signs of internal stress or external novelty in any high-risk environment can have disastrous consequences. Thus, at some point in time, for some individuals who have been adaptively competent in the past, cumulative experience may cease to be an asset and become associated with risk (108). The aircrew as well as the aircraft require a preflight check. Periodic reassessment of adaptive competence in individuals exposed to hazard is needed because the ability to adapt and respond appropriately varies within the context of immediate psychosocial circumstances as well as the psychophysiological state, and may change significantly with age *(103)*. Development of techniques combining neurometric and neuropsychological procedures appears to be the most appropriate direction to take for solving the problem of this reassessment.

Costs and Rewards of Isolated Duty

Socially nonadaptive individuals may be particularly disruptive to the efficiency of a small, closed group working in isolated conditions. Poorly socialized individuals may not be able to cope with the accelerated rate of self-disclosure and continual mutual exploration of personal value systems and goals which occur in isolation. Social competence is the ability to be aware of individual idiosyncracies and subtle interactions between group members that may be communicated both verbally and nonverbally. The socially competent individual adjusts his or her own behavior to be consistent and compatible with the microculture of the group. Open expression of discrepant personal values or aesthetic tastes may generate conflict within the group. Nonadaptive individuals presenting poorly defined, eccentric or culturally deviant value systems may have an especially difficult time maintaining their self-esteem under the continual pressure to conform which is present in an evolving group. These individuals may then respond with ineffective job performance, sulking, temper tantrums, consumatory behavior abuses with food, alcohol, etc., chronic complaints or self-imposed social isolation. The group may respond by using such a person as a scapegoat. One Antarctic veteran described life at a small isolated station as . . . "like being married." Thus, the person perceived as socially deficient or deviant may find himself faced with a surrogate multitude of nagging wives. There appears to be no taboo on explicit negative feedback in a small, closed group! In an effective, stable, coherent, well-functioning small group, the interactions may assume the special qualities of family living where manifestation of defensive self-images or affectation is not tolerated and there is little restraint of either affectionate or aggressive behavior.

It should be noted that the primary concern of the group during the acculturation process is to eliminate individual value systems that are not congruent with group values, norms, goals, and safety, and that threaten the group's ability to survive in an emergency. Once it has been established to the group's satisfaction that individuals are competent in their work, capable of dealing with potential problems, and actually assign group needs a priority over personal interests, then group concern with individual idiosyncratic behavior relaxes. Similar processes appear to be at work where peer selection techniques are used. Participation in and acceptance by such a group generates positive sentiments about a common social fate such as one finds among the graduates of private schools. In general, the veterans of isolated duty appear to have benefited from the acculturation experience and subsequent relaxation of conformity requirements by developing an ability to form a complex impression of other individuals which includes tolerance for more deviant behavior in others as well as in themselves. This makes it possible for them to interact with other people more effectively and may make their social encounters and relationships more enjoyable, effective, and creative.

The type of individual chosen for isolated duty is important in terms of personal health and well-being, individual job performance, and overall group effectiveness. Sells (109) suggested that the most socially adaptive individuals may be sensitive, mature, dedicated professionals whose desire for autonomy and emphasis on career goals takes precedence over being a husband, wife, father, mother, homeowner, and other common social roles. Individuals living in relatively confined quarters must be, or will learn to be, sensitive to subtle individual manifestations of territoriality, which may have temporal as well as spatial dimensions. They must also be able to contribute to minimizing environmental monotony by avoiding an indulgence in prolonged, fixed routines of dress, ritualistic behavior, and work scheduling. Conflict resolution during the structuring and homogenization of a new group is a significant source of stress. It is a poorly understood process, which is intensified and prolonged by the presence of socially inadequate or nonadaptive individuals. Individuals whose social-self is poorly developed and nonadaptive often rely upon direct social pressure from authority or force to regulate their behavior. They may find the resocialization process particularly unpleasant. These individuals are also unlikely to become productive members of the group, will be viewed as an additional burden in case of emergency, and may be prone to accidents, alcoholism, drug addiction, and symptoms associated with problems of psychological adjustment. The presence of such individuals may also provide a focus for polarization of the group into supporting and opposing factions. Such polarization produces an unhealthy environment for the entire crew.

The potential for nonadaptive behavior is greatly increased when a homogeneous group can not be established. Similar effects of the social environment upon mental health have been reported for rural communities by Leighton, Harding, Macklin, Macmillan, and Leighton *(110)*. These investigators found evidence for a strong correlation between social disintegration and the prevalence of psychiatric disorders. Crew selection has not yet been structured in terms of individual compatabilities and psychodynamic needs. Thus, at the small Antarctic station, no two years are ever exactly the same in terms of their influence upon the winter-over residents because of differences in the character, sentiment, and disposition of the microculture that evolves with each new winter-over crew.

The small Antarctic station is a total institution *(111)*. A total institution has been defined as a place of residence and work where a number of individuals lead an enclosed life cut off from wider society for an appreciable period of time. Long-term examples of total institutions include ships at sea, boarding schools, military academies, some university settings, military bases, work camps, prisons, and monasteries. The feature common to them all is that they *change* people--for some this is the primary task. Each is an experiment in what can be done to the self. A total institution established by society represents a formalized way in which individuals are constantly subjected to forces which effect planned changes in their characteristic modes of response to their environment. But many of the personality changes induced are often unplanned, uncontrolled consequences of other events or individual dispositions. The total institution achieves its goal by applying various stresses to the individual. Stresses related to physical isolation include isolation from the parent society, poor communication channels, crowding (with lack of space and privacy), various deprivations and, possibly, inadequate facilities for the use of leisure time. Social stresses appear to be the most important agents of change and these include: (a) devaluation of the self; (b) social deprivation; (c) status incongruency; (d) enforced socialization; (e) socio-cultural complexity; (f) inadequate opportunity for personal expression; (g) the development of identity conflicts; (h) cognitive dissonance; and (i) the constant threat of social rejection. At an Antarctic station all the inhabitants are subjected to some of these stresses at the outset. Some must struggle with them for the entire year.

At the small Antarctic station each man represents, in effect, a separate culture. By the end of the year, he is a member of a new culture somewhat different from any of those

originally represented. The problems and conflicts that
arise are only natural consequences of this resocialization
process. The greater the difference between the original
cultures represented, the greater the social stresses that
will develop and the greater the number of conflicts to be
resolved. Many of the initial attempts at conflict resolution are often irrational, systematically biased by incorrect judgments, or overly influenced by stereotypes so that
some groups may never achieve a stable configuration. Both
the crew's mission and the individual crewmembers suffer, in
ways described earlier, when there is no common culture
extensive enough and sensitive enough to regulate strong
countermotives, to promote task accomplishment, and to
harmonize social relationships.

Hartman *(112)* found that subjects isolated in a space
cabin simulator, on both solitary and two-man missions,
developed hostility that was directed toward the support
personnel outside the simulator. Hartman concluded that the
spaceflight environment develops a need for psychological
nurturance. Similar findings have been obtained in the
Antarctic *(48)*. External authority may come to be resented
because the need for it may not be obvious to the group and
actions requested may not satisfy the crew's conception of
what is proper or technically appropriate under the circumstances. The cultural drift caused by the resocialization
process is responsible for these problems to some extent.
Dependency and nurturing needs also generate hostility due
to what the crew interprets as undue delays in response to
requests for outside action or responses which reflect crew-
perceived incompetence, lack of sympathy, or poor perception
of the unique demands of their situation in isolation. The
more severe the isolation, the more likely it will be that
attempts at outside control will be resented, and the
greater will be the critical concern by the isolates over
the empathy of their outside superiors as well as their
competence in providing support.

Hartman *(112)* noted that his subjects showed meaningful
personality changes as a result of their short experience in
the simulator. The eight month Antarctic winter-over
experience may produce significant changes in an individual--both positive and negative. Evidently, any type of
personality change will involve some degree of behavior that
would normally be classified as indicative of mental illness
(113). The nonadaptive individual may have greater difficulty in recovering from this transition phase. Those
individuals who grow psychologically from isolated-duty
experiences tend to be more mature, tough-minded and
realistic, autonomous, and independent of authority *(89,114)*.

Acknowledgements

The reference services of the Hubertus Strughold Memorial Aeromedical Library of the USAF School of Aerospace Medicine, Brooks AFB, Texas, were of considerable assistance in the preparation of this report. However, the views expressed are those of the author and do not reflect any official policy or endorsement by the US Air Force. Drs. David Jones, Robert Ursano, and Richard Wheatley read portions of the manuscript during preparation and offered useful comments and criticisms. I am especially indebted to my editors, Stephen Cheston and David Winter, for the invitation to prepare this paper and for their patience, constructive comments, and continued encouragement during the task.

References and Notes

1. G.W. Barnard, Psychophysiology of Aerospace Medicine, In *Psychiatry in Aerospace Medicine*, C.J.G. Perry, Ed. (Little, Brown and Company, Boston, 1967), pp. 3-22.

2. P.M. Fine and B.O. Hartman, *Psychiatric Strengths and Weaknesses of Typical Air Force Pilots* (SAM-TR-68-121, USAF School of Aerospace Medicine, Brooks AFB, Texas, November 1968).

3. P.M. Fine and C.L. Jennings, *Coping and Developmental Theory: Applicability to the Selective Study of Normal Men* (USAF School of Aerospace Medicine, Aeromedical Review 1-65, Brooks AFB, Texas, April 1965).

4. G.R. Griffin and J.D. Mosko, *Naval Aviator Attrition 1950-1976: Implications for the Development of Future Research and Evaluation* (Naval Aerospace Medical Research Laboratory, Report No. 1237, Naval Air Station, Pensacola, Florida, August 1977).

5. R.A. McFarland, Psychological Factors in Selecting Aircrew, In *Human Factors in Air Transportation* (McGraw-Hill, New York, 1953).

6. A.L. Morgenstern, Emotional Suitability for a Flying Career, In *Psychiatry in Aerospace Medicine*, C.J.G. Perry, Ed. (Little, Brown and Company, Boston, 1967).

7. H. Hécaen and M.L. Albert, *Human Neuropsychology* (Wiley, New York, 1978).

8. M.D. Lezak, *Neuropsychological Assessment* (Oxford University Press, New York, 1976).

9. L. Small, *Neuropsychodiagnosis in Psychotherapy* (Brunner/Mazel, New York, 1973).

10. M.S. Gazzaniga and J.E. Le Doux, *The Integrated Mind* (Plenum Press, New York, 1978).

11. M. Calvin and O.G. Gazenko, Eds., *Foundations of Space Biology and Medicine*, 3 Vols. (NASA SP-374, U.S. Government Printing Office, Washington, D.C., 1975).

12. R.S. Johnson and L.F. Dietlein, Eds., *Biomedical Results from Skylab* (NASA SP-377, U.S. Government Printing Office, Washington, D.C., 1977).

13. M. Collins, Carrying the Fire, In *Skywriting: An Aviation Anthology*, J. Gilbert, Ed. (St. Martins Press, 1978).

14. S.G. Rosen, "Mind in Space," *USAF Medical Service Digest* 27 (1976), pp. 4-17.

15. Space Science Board, *Human Factors in Long-Duration Space Flight* (National Academy of Sciences, Washington, D.C., 1972).

16. C.J.G. Perry, "Psychiatric Selection of Candidates for Space Missions," *Journal of the American Medical Association 194* 8 (1965), pp. 99-102.

17. C.J.G. Perry, "A Psychiatric 'Back-up' System for Selection of Space Crews," *Amer. J. Psychiatry 123* (1967), pp. 821-825.

18. C.J.G. Perry, Psychiatric Support for Man in Space, In *Psychiatry in Aerospace Medicine*, C.J.G. Perry, Ed. (Little, Brown and Company, Boston, 1967).

19. K. Natani, "A Voice from the Field: Comment on the Role of the Psychologist in the Peace Corps," *Amer. Psychologist 29* (1974), pp. 59-63.

20. E.K.E. Gunderson, Psychological Studies in Antarctica, In *Human Adaptability to Antarctic Conditions*, E.K.E. Gunderson, Ed. (American Geophysical Union, Antarctic Research Series, Volume 22, Washington, D.C., 1974).

21. S.B. Sells and E.K.E. Gunderson, A Social System Approach to the Long-Duration Space Mission, In Space Science Board, *Human Factors in Long-Duration Space Flight* (National Academy of Sciences, Washington, D.C., 1972).

22. A.P. Avtsyn, Ed., *Human Acclimatization in the Polar Regions* [Russian, 1969] (National Technical Information Service Translation, Report No. 56252, Arlington, Va., 1972).

23. V.A. Bugaev, Ed., *Soviet Antarctic Research 1956-1966* [Russian, 1967] (National Technical Information Service Translation, Report TT 69-55004, Arlington, Va., 1972).

24. O.G. Edholm and E.K.E. Gunderson, Eds., *Polar Human Biology* (Wm. Heinemann Medical Books Ltd., London, 1973).

25. Ye. S. Korotkevich, Ed., *Soviet Antarctic Expedition Information Bulletin 7* 74 [Russian, 1969] (American Geophysical Union Translation, Washington, D.C., 1971).

26. A.L. Matusov, Ed., *Medical Research on Arctic and Antarctic Expeditions*, Volume 299 [Russian, 1971] (National Technical Information Service Translation, Report TT 72-50004, Arlington, Va., 1973).

27. O. Wilson, Human Adaptation to Life in Antarctica, In *Biogeography and Ecology in Antarctica*, J. Van Meighen, P. Van Oye, J. Schell, Eds. (Monographie Biologicae, Volume 15, W. Junk, The Hague, Netherlands, 1965).

28. E. Stuhlinger, Operator Requirements, In *Psychophysiological Aspects of Space Flight*, B.E. Flaherty, Ed. (Columbia University Press, New York, 1961).

29. E. Stuhlinger, personal communication, Amundsen-Scott Station, Antarctica, 8 January 1967, In K. Natani, *South Pole Diary* (unpublished) 1967, p. 31.

30. W.M. Smith and M.B. Jones, "Astronauts, Antarctic Scientists and Personal Autonomy," *Aerospace Medicine 33* (1962), pp. 162-166.

31. B.O. Hartman and D.E. Flinn, Crew Structure in Future Space Missions, In *Lectures in Aerospace Medicine* (USAF School of Aerospace Medicine, Brooks AFB, Texas, February, 1964).

32. J.T. Shurley, K. Natani, R. Sengel, Ecopsychiatric Aspects of a First Human Space Colony, In *Space Manufacturing Facilities II (Space Colonies)*, J. Grey, Ed. (American Institute of Aeronautics & Astronautics, New York, 1977).

33. R. Radloff, Naturalistic Observations of Isolated Experimental Groups in Field Settings, In *Man in Isolation and Confinement*, J.E. Rasmussen, Ed. (Aldine, Chicago, 1973).

34. D.G. Simons, B.W. Henderson, J.L. Riehl, Personal Experiences in Space Equivalent Flight, In *Psychophysiological Aspects of Space Flight*, B.E. Flaherty, Ed. (Columbia University Press, New York, 1961).

35. R. Radloff, Research on Life and Work in Undersea Habitats, In *Behavior, Design, and Policy Aspects of Human Habitats*, W.M. Smith, Ed. (Office of Community Outreach, University of Wisconsin, Green Bay, 1972).

36. K. Natani, Observations of Mood and Performance in a Small Group of Men During Eight Months of Isolated Duty in Antarctica, unpublished thesis, University of Oklahoma (December, 1971).

37. Eason and Harter *(115)* hypothesized that certain changes in sensory and perceptual processes such as exaggerated orienting responses, changes in color discrimination, development of behavioral apathy, and the occurrence of hallucinations might be found in long-duration space missions. My personal observations during an Antarctic winter confirm the occurrence of all these phenomena in varying degrees in one or more of the men. Hallucinatory activity may be very subtle. Such illusory perceptions require immediate verification by other members of the crew, and demonstration of their hallucinatory basis may require several days of standing watch with the individual who may be experiencing these phenomena.

38. J.T. Shurley, "Man on the South Polar Plateau," *Archives of Internal Medicine 125* (1970), pp. 625-629.

39. J.T. Shurley, New Directions in Antarctic Biomedical Research, In *Research in Antarctica*, L.O. Quam, Ed. (American Association for the Advancement of Science, Washington, D.C., 1971).

40. J.T. Shurley, Antarctica is Also a Prime Natural Laboratory for the Behavioral Sciences, In *Polar Human Biology*, O.G. Edholm and E.K.E. Gunderson, Eds. (Wm. Heinemann Medical Books, London Ltd., 1973).

41. J.T. Shurley, Physiological Research at United States Stations in Antarctica, In *Human Adaptability to Antarctic Conditions*, E.K.E. Gunderson, Ed. (American Geophysical Union, Antarctic Research Series, Volume 22, Washington, D.C., 1974).

42. G.M. Danishevskii, V.N. Ponomarev, I.I. Tikhomirov, Human Acclimatization in the Antarctic, In *Soviet Antarctic Research 1956-1966*, V.A. Bugaev, Ed. [Russian, 1967] (National Technical Information Service Translation, Report TT 69-55004, Arlington, Va., 1970).

43. Hypokinesia is a significant reduction in the activity of the large muscles of the body, primarily the muscles used to working against the force of gravity. Rest in bed is an example of a hypokinetic state. Conditions both inside and outside Antarctic stations tend to encourage sedentary activities and many of the residents may spend much of their time sitting, lounging, or resting in bed. This is especially true in the winter, when protective clothing is required that greatly restricts one's ability to move rapidly and efficiently. Evidence available from Antarctic studies suggests that this prolonged reduction in activity may have cumulative effects similar to those seen in prolonged bedrest or weightlessness. This is an area that requires further study.

44. S. Smith, Studies of Groups in Confinement, In *Sensory Deprivation: Fifteen Years of Research*, J.P. Zubek, Ed. (Appleton-Century-Crofts, New York, 1969).

45. W.H. Teichner, "An Exploration of Some Behavioral Techniques for Toxicity Testing," *J. Psychology 65* (1967), pp. 69-90.

46. B.O. Hartman, Time and Load Factors in Astronaut Proficiency, In *Psychophysiological Aspects of Space Flight*, B.E. Flaherty, Ed. (Columbia University Press, New York, 1961).

47. B.O. Hartman, Psychological Factors in Flying Fatigue, In *Psychiatry in Aerospace Medicine*, C.J.G. Perry, Ed. (Little, Brown and Company, Boston, 1967).

48. K. Natani and J.T. Shurley, Sociopsychological Aspects of a Winter Vigil at South Pole Station, In *Human Adaptability to Antarctic Conditions*, E.K.E. Gunderson, Ed. (American Geophysical Union, Antarctic Research Series, Volume 22, Washington, D.C., 1974).

49. A.B. Blackburn, J.T. Shurley, K. Natani, Psychological Adjustment at a Small Antarctic Station: An MMPI Study, In *Polar Human Biology*, O.G. Edholm and E.K.E. Gunderson, Eds. (W. Heinemann Medical Books, London, 1973), pp. 369-383.

50. K. Natani, J.T. Shurley, C.M. Pierce, R.E. Brooks, "Long-Term Changes in Sleep Patterns in Men on the South Polar Plateau," *Archives of Internal Medicine 125* (1970), pp. 655-659.

51. K. Natani and J.T. Shurley, Extrinsic Parameters and the Self-Regulation of Sleep in Antarctica, In *Self-Regulation of the Sleep Processes*, N.I. Moiseeva, Ed. (Academy of Sciences of the USSR [In Russian], 1977).

52. K. Natani and J.T. Shurley, "Laterality Effects in Adaptation," Paper prepared for inclusion in the proceedings of the Fourth International Symposium on Circumpolar Health (Novosibirsk, USSR, October 1978).

53. R.B. Livingston, Psychological and Neuromuscular Problems Arising from Prolonged Inactivity, In *Human Ecology in Space Flight II*, D.H. Calloway, Ed. (New York Academy of Sciences, New York, 1967)

54. The western approach places primary emphasis upon a taxonomy of operator skills or abilities and task characteristics. The work of Fitts and Fleishman is representative of efforts in this area. Fleishman *(116, 117)* has consistently emphasized individual differences in abilities, and has recently addressed the problem of defining and classifying tasks in terms of the abilities required to perform them *(118)*. A taxonomy for the classification of tasks and operator requirements in terms of their interrelationships may be extremely useful in estimating the psychophysiological state of the individual under different conditions, as well as in determining the types of tasks which human operators should not be expected to perform well under stress.

    The taxonomic approach assumes that high levels of experience with an overlearned task decrease individual differences and make performance resistant to stress effects *(119)*. Thus, environmental stressors are expected to produce specific, predictable performance decrements in individual tasks. Data from the Antarctic and other stress studies suggest that, in addition, the possibility of irrational behavior is an important consideration and that the emotional state of the operator during stressed task performance may be important in terms of selecting competent individuals and in predicting operator reliability under adverse conditions.

55. B.K. Lester, N.R. Burch, R.C. Dossett, "Nocturnal EEG-GSR Profiles: The Influence of Presleep States," *Psychophysiology 3* (1967), pp. 238-248.

56. B.O. Hartman, "Field Study of Transport Aircrew Workload and Rest," *Aerospace Medicine 42* (1961), pp. 817-821.

57. D.A. Harris, G.V. Pegram, B.O. Hartman, "Performance and Fatigue in Experimental Double-Crew Transport Missions," *Aerospace Medicine 42* (1971), pp. 980-986.

58. D.M. Denison, F. Ledwith, E.C. Poulton, "Complex Reaction Times at Simulated Cabin Atltitudes of 5,000 Feet and 8,000 Feet," *Aerospace Medicine 37* (1966), pp. 1010-1013.

59. J. Ernsting, "Prevention of Hypoxia - Acceptable Compromises," *Aviation, Space, and Environmental Medicine, 49* (1978), pp. 495-502.

60. R.A. McFarland, Psychophysiological Implications of Life at Altitude and Including the Role of Oxygen in the Process of Aging, In *Physiological Adaptations*, M.K. Yousef, S.M. Horvath, R.W. Bullard, Eds. (Academic Press, New York, 1972).

61. W.H. Teichner and D.E. Olson, Predicting Human Performance in Space Environments (NASA CR-1370, National Technical Information Service, Springfield, Va., 1969).

62. G.S. Tune, "Psychological Effects of Hypoxia: Review of Certain Literature from the Period 1950-1963," *Perceptual and Motor Skills 19* (1964), pp. 551-561.

63. F. Ledwith, "The Effects of Hypoxia on Choice Reaction Time," *Ergonomics 13* (1970), pp. 465-482.

64. Results of these recent studies of the influence of mild hypoxic stress upon learning have suggested that the aircraft pressurization standard for aircrews in routine flight should be changed from 8,000 feet to 6,000 feet. Ernsting *(59)* has concluded that . . . "the mild hypoxia produced by breathing air at an altitude of 8,000 ft should not be accepted for aircrew engaged in air operations because of the very significant impairment of ability to respond to a novel complex situation which it induces".

65. In a personal communication with the author, Teichner noted that the most interesting subjects, from the standpoint of providing responses that support the presence of significant stress effects, were usually the ones who dropped out of experiments. These were often individuals experiencing high levels of psychological stress in their lives outside the laboratory. This was also a problem in the Antarctic. Behavioral observations indicated that the Seabee group suffered

the greatest stress-related effects, but we were able
to obtain only one of these individuals as a volunteer
for the sleep studies. Nevertheless, an advantage of
the field study, where one often has a captive group,
is that the behavior of these individuals can be
observed, and that they may eventually volunteer to
become members of the experimental group.

66. D.E. Flinn, B.O. Hartman, D.H. Powell, R.E. McKenzie,
Psychiatric and Psychological Evaluation, In *Aeromedical Evaluation for Space Pilots*, L.E. Lamb, Ed.
(USAF School of Aerospace Medicine Special Report,
Brooks AFB, Texas, July, 1963).

67. W.H. Teichner, "Interaction of Behavioral and Physiological Stress Reactions," *Psych. Rev.*, *75* (1968),
pp. 271-291.

68. W.H. Teichner and D.E. Olson, "A Preliminary Theory of
the Effects of Task and Environmental Factors on Human
Performance," *Human Factors 13* (1971), pp. 295-344.

69. Research in the use of multiple input-multiple response
tasks for pilot selection is now being actively pursued
by the Navy at Pensacola, Florida *(120,121,122)*. This
research would not be possible without the use of
computer technology *(123)*. The work represents a significant change in Navy policy. The Navy had previously
relied exclusively upon psychometric measures for pilot
selection, but high attrition levels have stimulated
this interest in performance testing procedures *(4)*.

70. A group of researchers in the Department of Psychiatry
and Behavioral Sciences at the University of Oklahoma
Health Sciences Center, under the direction of Oscar
Parsons, constructed a special apparatus for lateralized perimetric (as opposed to the standard central
binocular) presentation of flickering light to a
subject for determining flicker fusion rates (CFF).
Investigations using this modified CFF technique
suggested that flicker perimetry provides a sensitive
technique for the detection of neuropsychological
deficits related to cerebral lesions *(124,125,126,127,
128)*. The standard central presentation technique has
been discarded by many clinicians because of low
sensitivity.

It is now generally accepted that each cerebral
hemisphere in the mammalian brain constitutes an integrated information processing system possessing highly

specialized functions *(129)*. In the human brain, for
most individuals, the left hemisphere performs language-
and speech-related sequential, logical processes under
conscious control. The right hemisphere, on the other
hand, evidently controls parallel, spatially related
semi-conscious or automatic processes. Some aspects of
this lateral differentiation appear to be present in
the newborn, but recent work *(130,131)* suggests that
the functional differentiation and development of
cerebral dominance is a life-span developmental process.

71. K. Natani, Laterality Effects in a Tachistoscopic Optional Shift Task in Young Adults, Unpublished thesis, University of Oklahoma Health Sciences Center, Oklahoma City. *Dissertation Abstracts International* 38 (1977), p. 744.

72. K. Natani, E. Engelken, D. Threatt, "PDP-12 Stimulus Control and Data Collection for Lateralized Discrimination Learning," *Behavior Research Methods & Instrumentation* 11 (1979), pp. 264-270.

73. Bandwidth is a term related to filtering processes in the physical sciences. A narrow-bandwidth filter is a highly selective device which can be made or set to pass only one small portion of a sound, radio, or lightwave signal. The application of this concept to information processing in the nervous system was first suggested by Broadbent *(132)*. He conceptualized the process of attention as a filtering system with variable bandwidth properties. A narrow bandwidth setting is efficient for processing information received in a single stream or sequence, such as speech or a line of print. A wide bandwidth setting is efficient for processing information with much greater variety and diversity, such as music or an illustration. The lateralization conceptualization of hemispheric specialization suggests that as one listens to speech the narrowband left hemisphere processes focus upon the overt information content of the speech received, while wideband right hemisphere processes deal with features of the context in which the speech is produced and delivered. The two processes may occur simultaneously, but integrating both types of information to produce a response may require a measurable amount of time. When an inappropriate filter (hemisphere) is applied to a given information processing task one would expect a loss in processing efficiency and a proportionate increase in processing time.

74. Activation level refers to the state of psychophysiological arousal existing in an individual, e.g., relaxed or excited. Activation refers to the internal state of an organism and appears to influence any type of response an individual may produce. It has been shown that activation, as measured by various psychophysiological indices, e.g., heart rate, skin conductance, pupil size, etc., is related to performance. Optimum performance is related to intermediate levels of activation, thus producing a correlation illustrated by an inverted U-shaped curve. Activation levels that are too low or too high interfere with efficient performance. Duffy *(133)* has provided an excellent and extensive review of this concept.

75. K. Natani, "Letters: the Split and Whole Brain, A Comment on Body Position and Thinking," *Human Nature 1* (1978), p. 9 & 96.

76. J.E. Bogen, "The Other Side of the Brain II: An Appositional Mind," *Bull. Los Angeles Neurological Soc. 34* (1969), pp. 135-162.

77. J. Levy and C. Trevarthen, Metacontrol of Hemispheric Functioning in Human Split Brain Patients," *J. Exper. Psych.: Human Perception and Performance 2* (1976), pp. 299-312.

78. Dichotic listening appears to be a procedure useful for the measurement of base rates in hemispheric alternation. In a dichotic listening task, different sounds are presented simultaneously to each ear. Response time is usually limited, with fast responding emphasized. There are several variations on the procedure: (a) the subject is to determine to which ear a particular sound was presented; (b) the subject must indicate all sounds presented to both ears; or (c) the subject must respond to all sounds presented to one ear and only respond to certain designated sounds presented to the opposite ear. Under these conditions a block is defined as a missed response opportunity for both ears or an incorrect response for both ears. A missed or incorrect response for only one ear indicates that the opposite ear had an information processing advantage on that trial.

79. G. Lazarus-Mainka, and H. Horman, "Strategic Selection (Metacontrol) of Hemisphere Dominance in Normal Human Subjects," *Psych. Res. 40* (1978), pp. 15-25.

80. There is now evidence to support the influence of experience upon the development of lateral differentiation in the brain of the rat *(134)*, and studies of the human brain suggest that there may be a differential biochemical substrate for lateralization of function *(135)*.

81. H.H. Jasper, "A Laboratory Study of Diagnostic Indices of Bilateral Neuromuscular Organization in Stutterers and Normal Speakers," *Psychological Monographs 43* (1932), pp. 72-174.

82. R.W. White, Strategies of Adaptation: An Attempt at Systematic Description, In *Coping and Adaptation*, G.V. Coelho, D.A. Hamburg, J.E. Adams, Eds. (Basic Books, New York, 1974).

83. J.D. Matarazzo, *Wechsler's Measurement and Appraisal of Adult Intelligence* (Williams & Wilkins, Baltimore, 1972).

84. R.W. White, "Motivation Reconsidered: The Concept of Competence," *Psych. Rev. 66* (1959), pp. 297-333.

85. R.W. White, "Ego and Reality in Psychoanalytic Theory," *Psych. Issues, 3* (1963), Monograph 11, pp. 210 (International Universities Press, New York).

86. R.S. Ezekiel, "The Personal Future and Peace Corps Competence," *J. Personality and Soc. Psych.*, Monograph Supplement, *8*, Part 2 (1968), pp. 1-26.

87. R.E. Jensen, The Concept of Competence: A Provisional Attempt at Construct Validation," JSAS MS. No. 559, Abstracted in the *JSAS Catalog of Selected Documents in Psychology 4* 14 (1974).

88. F.B. Tyler, "Individual Psychosocial Competence: A Personality Configuration," *Educational and Psychological Measurement 38* (1978), pp. 309-323.

89. M.B. Smith, "Explorations in Competence: A Study of Peace Corps Teachers in Ghana," *Amer. Psychologist 21* (1966), pp. 555-566.

90. N.D. Sundberg, L.R. Snowden, W.M. Reynolds, "Toward Assessment of Personal Competence and Incompetence in Life Situations," *Ann. Rev. Psychology 29* (1978), pp. 179-221.

91. P.M. Fine and C.L. Jennings, "Personality Development: Applications of Theory to Problems of Aerospace Selection," *Aerospace Medicine 37* (1966), pp. 695-701.

92. G.E. Vaillant, *Adaptation to Life* (Little, Brown and Company, Boston, 1977).

93. R.E. Reinhardt, "The Outstanding Jet Pilot," *Amer. J. Psychiatry 127* (1970), pp. 32-36.

94. D. Mechanic, Social Structure and Personal Adaptation: Some Neglected Dimensions, In *Coping and Adaptation*, G.V. Coelho, D.A. Hamburg, and J.E. Adams, Eds. (Basic Books, New York, 1974).

95. D.A. Hills, College-Based Mental Health Units and the Selection of ROTC Candidates for Future Flight Training, In *Psychiatry in Aerospace Medicine*, C.J.G. Perry, Ed. (Little, Brown and Company, Boston, 1967).

96. W.M. Smith, "Scientific Personnel in Antarctica: Their Recruitment, Selection and Performance," *Psychological Reports 9* (1961), pp. 163-182.

97. M.S. Carpenter, Sea Lab II Crew Selection, Training, and Daily Operations, In *Man's Extension Into the Sea*, Transactions of a joint symposium, 11-12 January 1966, Washington, D.C. (National Technical Information Service, Springfield, Va., 1966), pp. 55-60.

98. D.E. Flinn, Psychiatric Factors in Astronaut Selection, In *Psychophysiological Aspects of Space Flight*, B.E. Flaherty, Ed. (Columbia University Press, New York, 1961).

99. M. Zuckerman, The Sensation Seeking Motive, In *Progress in Experimental Personality Research*, Volume 7, B.A. Maher, Ed. (Academic Press, New York, 1974).

100. A.L. Morgenstern, Phobic Reactions to Flying, In *Psychiatry in Aerospace Medicine*, C.J.G. Perry, Ed. (Little, Brown and Company, Boston, 1967).

101. J.E. Nardini, R.S. Hermann, J.E. Rasmussen, "Navy Psychiatric Assessment Program in the Antarctic," *Amer. J. Psychiatry 119* (1962), pp. 97-105.

102. W.F. Belk, "Psychophysiologic Factors in USAF Aircraft Accidents, 1974-1977," Paper presented at the Ninth International Society of Air Safety Investigators annual meeting, Seattle, Washington, 3-5 October 1978.

103. S.J. Gerathewohl, *Psychophysiological Effects of Aging--Developing a Functional Age Index for Pilots: II-Taxonomy of Psychological Factors* (FAA Office of Aviation Medicine, Report No. FAA-AM-78-16, Washington, D.C., March 1978).

104. D.S. Ricketson, S.A. Hohnson, L.B. Branham, R.K. Dean, Incidence, Cost and Factor Analysis of Pilot-Error Accidents in U.S. Army Aviation, *Proceedings of Specialists' Meeting on Behavioral Aspects of Aircraft Accidents*, Soesterberg, Netherlands, 3-7 September 1973 (NATO Advisory Group for Aerospace Research and Development, 1973).

105. B. Brechner, "Aftermath Tenerife," *Flying Magazine 104* (1979), pp. 23-26.

106. D. Noland, "The Anatomy of a Crunch," *Air Progress 30* (1973), pp. 26-29.

107. Editors of *Flying Magazine, Pilot Error: Anatomies of Aircraft Accidents* (Van Nostrand Reinhold, New York, 1977).

108. C.F. Booze, Jr., *An Epidemiological Investigation of Occupation, Age and Exposure in General Aviation Accidents* (FAA Office of Aviation Medicine Report No. FAA-AM-77-10, Washington, D.C., March 1977).

109. S.B. Sells, The Taxonomy of Man in Enclosed Space, In *Man in Isolation and Confinement*, J.E. Rasmussen, Ed. (Aldine, Chicago, 1973).

110. D.C. Leighton, J.S. Harding, D.B. Macklin, A.M. Macmillian, A.H. Leighton, *The Character of Danger: Psychiatric Symptoms in Selected Communities* (Basic Books, New York, 1963).

111. E. Goffman, *Asylums: Essays on the Social Situation of Mental Patients and Other Inmates* (Doubleday, New York, 1961).

112. B.O. Hartman, Experimental Approaches to the Psychophysiological Problems of Manned Space Flight, In

Lectures in Aerospace Medicine (USAF School of Aerospace Medicine, Brooks AFB, Texas, January, 1961).

113. E.A. Haggard, Isolation and Personality, In *Personality Change*, P. Worchel and D. Byrne, Eds. (Wiley, New York, 1964).

114. A.J.W. Taylor and J.T. Shurley, "Some Antarctic Troglodytes," *Internat. Rev. Applied Psychology 20* (1971), pp. 143-148.

115. R.G. Eason and M.R. Harten, Sensory, Perceptual and Motor Factors, In *Human Factors in Long-Duration Space Flight*, Space Science Board (National Academy of Sciences, Washington, D.C., 1972).

116. E.A. Fleishman, "Testing for Psychomotor Abilities by Means of Apparatus Tests," *Psych. Bull. 50* (1953), pp. 241-262.

117. E.A. Fleishman, Individual Differences in Motor Learning, In *Learning and Individual Differences*, R.M. Gagne, Ed. (Charles E. Merrill Books, Columbus, Ohio, 1967).

118. E.A. Fleishman, "Relating Individual Differences to the Dimensions of Human Tasks," *Ergonomics 21* (1978), pp. 1007-1019.

119. P.M. Fitts, Factors in Complex Skill Training, In *Training Research and Education*, R. Glaser, Ed. (University of Pittsburgh Press, Pittsburgh, 1962).

120. D. Gopher and D. Kahneman, "Individual Differences in Attention and the Prediction of Flight Criteria," *Perceptual and Motor Skills 33* (1971), pp. 1335-1342.

121. R.A. North and D. Gopher, "Measures of Attention as Predictors of Flight Performance," *Human Factors 18* (1976), pp. 1-14.

122. R.A. North and G.R. Griffin, *Aviator Selection 1919-2977* (Naval Aerospace Medical Research Laboratory, Special Report 77-2, Naval Air Station, Pensacola, Florida, October, 1977).

123. R.S. Gibson, Computer Measurement of Complex Performance, In *Medical Requirements and Examination Proce-*

dures in Relation to Tasks of Today's Aircrew: Comparison of Examination Techniques in Neurology, Psychiatry, and Psychology With Special Emphasis on Objective Methods and Assessment Criteria, H. Oberholz, Ed. (AGARD Conference Proceedings No. 153, North Atlantic Treaty Organization, September 1974).

124. O.A. Parsons and M.M. Huse, "Impairment of Flicker Discrimination in Brain-Damaged Patients," *Neurology 8* (1958), pp. 750-755.

125. O.A. Parsons, P.J. Chandler, R.W. Teed, G. Haase, "Comparison of Flicker Perimetry and Standard Visual Fields in Brain-Damaged Patients," *Acta Neurologica Scandinavia 42* (1966), pp. 207-212.

126. O.A. Parsons, R.K. Majumder, P.J. Chandler, "Impaired Flicker Detection in Visual Fields Subserved by Non-Damaged Hemispheres," *Cortex 3* (1967), pp. 307-316.

127. O.A. Parsons, J. Burn, P.J. Chandler, "Cerebral Dysfunction and Flicker Detection: The Role of Local Adaptation," *Amer. J. Psychology 81* (1968), pp. 525-534.

128. O.A. Parsons, R.K. Majumder, P.J. Chandler, "Brain Damage, Light-to-Cycle Ratio and Flicker Detection," *Cortex 4* (1968), pp. 269-279.

129. B. Milner, Hemisphere Specialization: Scope and Limits, In *Neurosciences Third Study Program*, F.O. Schmitt and F.G. Worden, Eds. (MIT Press, Cambridge, MA, 1974).

130. J.W. Brown and J. Jaffe, "Hypothesis on Cerebral Dominance," *Neuropsychologia 13* (1975), pp. 107-110.

131. A.R. Luria, Neuropsychology: Its Sources, Principles and Prospects, In *The Neurosciences: Paths of Discovery*, F.G. Worden, J.P. Swazey and G. Adelman, Eds. (MIT Press, Cambridge, MA, 1975).

132. D.E. Broadbent, *Perception and Communication* (Pergamon Press, London, 1958).

133. E. Duffy, Activation, In *Handbook of Psychophysiology*, N.S. Greenfield and R.A. Sternback, Eds. (Holt, Rinehart and Winston, New York, 1972), pp. 577-622.

134. V.H. Denenberg, J. Garbanati, G. Sherman, D.A. Yutzey, R. Kaplan, "Infantile Stimulation Induces Brain Lateralization in Rats," *Science 201* (1978), pp. 1150-1151.

135. A. Oke, R. Keller, I. Mefford, R.N. Adams, "Lateralization of Norepinephrine in Human Thalamus," *Science 200* (1978), pp. 1411-1413.

_____ *Joan E. Sieber*

3. Well-Being and Privacy in Space: Anticipating Conflicts of Interest

<u>Introduction</u>

Each of us has an idea of what we mean by *our privacy*, and from time to time we may complain that we do not have enough, or else that we have too much of it, or that someone has intolerably invaded our privacy. Yet, when we try to say what we mean by privacy, or when we try to define what is private to someone else, we discover that the concept is difficult to define. Our ideas of privacy and our accustomed ways of protecting our privacy are deeply embedded in our rules of etiquette, in the architecture of our buildings and automobiles and landscape, in our verbal and nonverbal means of communicating, in our role relationships and the degree of power we have in them, and in our knowledge that constitutional law guarantees us certain forms of privacy. We need only move to another culture, or to another subculture within our own society, to have it forcibly brought home to us that privacy is relative to many other factors, and that the loss of accustomed forms of privacy is a form of culture shock that reduces our satisfaction and effectiveness in life.

When humans move into outer space for purposes of science or industry, or to establish a permanent habitat, the loss of accustomed privacies and the creation of new forms of privacy are sure to be major concerns. And, just as the concept of privacy has grown immensely in salience during the past hundred years, and especially in the last decade, it is fair to predict that people's needs for and claims to privacy in space habitats will grow and change in relation to the other evolving characteristics of culture and physical environment in space.

The purpose of this article is to introduce the concept of privacy in a way that will (a) increase awareness of the difficulties we may anticipate in providing a decent respect for privacy to those who are among the first to inhabit space, (b) increase sensitivity to the factors that influence the need for privacy, and (c) increase awareness of the importance of anticipating the privacy needs of early space inhabitants. In regard to the latter hope, it will become evident in the course of the discussion that it is not the role of earth dwellers now to anticipate the privacy needs of residents of well-established space communities of the future. Privacy is so relative to other aspects of one's experience, personal development, culture, and physical environment that present-day earth dwellers would be naive to assume that they could accurately anticipate those conditions and those needs. It will then behoove us to be sensitive to the relativity of privacy and not to expect irrelevant or earth-relevant protections to privacy to serve the needs of persons for whom privacy means something else.

In this article, first, the relativity of the notion of privacy is introduced. Second, the relationship is indicated between lack of privacy--however the individual defines this--and stress that leads to degradation of health and performance. Finally, a few problems of privacy that early space inhabitants may encounter are outlined and suggestions are offered for ameliorating these problems.

Some definitions of privacy that are commonly given in our society are presented and criticized, to show that privacy means so many things to middle-class Americans that it is difficult to specify the defining attributes of privacy even for this limited population. Privacy is then examined in a broader cultural and psychological context. A theory of privacy is introduced that sets forth factors that affect people's subjective definition of privacy and their ways of protecting their privacy. It will become apparent that we can educate ourselves to become more sensitive to the privacy needs of others, but that we cannot know for certain how others will want to regulate their privacy, particularly if their life circumstances differ from our own. It will also become apparent that there needs to be sensitive communication between persons about their individual needs for privacy and about the development of mutual arrangements for respecting these needs.

The Relativity of Privacy

Privacy is often defined as the individual's "right to be let alone" *(1)*. Alternatively, Westin has defined privacy as "the claim of individuals, groups or institutions to determine for themselves when, how and to what extent information about them is communicated to others" *(2)*. However, as privacy is defined in American society, it means many more things and different things than these definitions include. (In another sense it is also less inclusive than these definitions would imply, since interests in privacy often conflict with other legitimate interests, and society is not obliged to "let [persons] alone" under all circumstances. The limitations of privacy are discussed elsewhere *(3)*.)

As Greenawalt *(4)* has pointed out, privacy is not a right, it is a situation--a state of being apart, physically and emotionally, from others. One may then have needs or claim rights relative to one's situation of privacy. For example, space travelers might experience too much privacy--too little opportunity to share thoughts and experiences with cherished others, or too little opportunity to be observed or to observe others; thus, they might claim a need or a right to travel in the company of friends. Space travelers might also experience too little privacy--too little opportunity to be alone and develop and nurture original ideas, too little opportunity to perform personal acts free from the scrutiny of others, and too little opportunity to control what becomes public information about their activities, health, or fate. One can readily imagine that space travelers will experience simultaneously a need for less privacy of some kinds and for more of other kinds. For example, an intolerable loneliness for friendship might be coupled with the intolerable breach of privacy contained in having biomedical data about themselves continuously telemetered to earth and given to the news media to interpret in ways that will titillate the public. These conditions might motivate space travelers to assert various rights to privacy--some having to do with satisfaction of needs for personal sharing and others having to do with satisfaction of needs to control the flow of personal information to strangers.

Privacy is not simply control of information about oneself. It is also control of information that others offer us and control of our own thought processes. Thus, privacy is invaded when persons are exposed to unwanted information about others. For example, space travelers

might find that their co-travelers breach privacy with
discussions they would rather not hear and with activities
they would rather not see. Space travelers might also
claim that their thought processes--their most private
element--are invaded and controlled by the limiting of
information to them or by the drugs they are given to
suppress motion sickness.

Many notions of privacy have been introduced, but
none of them is particularly foreign to us; all of these
are notions that middle-class American adults can
readily understand. To expand this notion, let us now
look at some examples of privacy regulation in other
cultures.

The Balinese culture is characterized by great privacy
in relation to those who are not close friends or members
of one's family. Homes are surrounded by high walls,
and entrance ways are generally not open to outsiders.
Yet within these enclaves there is "a tremendous warmth,
humor, [and] openness" among friends and family (5).

In contrast, there are communities in Java where there
is little physical privacy. There are no fences and few
doors, the flimsy, thin-walled bamboo houses are open to
anyone, and people tend to wander unannounced into all
parts of a house. But social interaction in these
communities is formal and minimal. Individuals hide their
emotions, speak softly, engage in an elaborate social
decorum, and "shut people out with a wall of etiquette" (6).

The Mehinacu Indians of Brazil also have little
physical privacy. They live in communal houses and are
gregarious, so that people generally know what the other
inhabitants are doing. However, it is customary not to
expose the misconduct of another or to ask embarrassing
questions, and lying is regularly used to avoid revealing
private information. For a member of this tribe there
are times in one's life when it is the custom to live in
isolation, and there are places in the forest where one
can escape to privacy when that seems desirable (7).

As these examples illustrate, what is a meaningful
or acceptable way to have and preserve privacy in one
culture may be meaningless or unacceptable in another.
Even within our own society, there are many individual
differences in the meaning and significance of privacy.
Laufer and Wolfe (8) have presented a developmental theory
of privacy that employs the dimensions of *self-ego,
environmental, interpersonal,* and *control choice* factors

to take into account the kinds of variables that produce cultural and individual differences in the meaning of privacy and the need for a means of protecting it. That theory of privacy is summarized as follows:

The *self-ego* dimension refers to the development of autonomy and personal dignity. In infancy and early childhood, aloneness is aversive. But within a few years aloneness becomes a frequently sought state in which the individual can establish a sense of self and autonomy and can develop, nurture, and protect new ideas, thereby creating a basis for self-esteem, personal strength, and dignity. The adult continues to need periods of aloneness, but adults with a well-developed sense of autonomy and dignity are able to protect their privacy through various other means when physical solitude is not available.

The *environmental* dimension includes (1) cultural, (2) sociophysical, and (3) life-cycle elements. *Cultural* elements (1) include the norms that are transmitted through tradition, language, and values. Each culture transmits only certain particular means of protecting privacy, thus limiting the options the individual perceives as available for being private; e.g., one culture may permit lying, while another may permit persons to have private rooms in which to spend a major portion of each day. Sociophysical elements (2) refer to the physical settings one has available and to the social factors that determine the use of those settings. Thus, indoor bathrooms offer a private place for members of small families, but not for members of large, poor families who have only one bathroom. Places where group members can seek solitude in zero gravity may not be places that are available in a gravitational field (consider the members of groups that interpret hanging in corners to be a means of rejecting the group). Life-cycle elements (3) are the ever-changing elements in one's life that vary with age, occupation, available technology, and changing sociocultural patterns. The privacies that one needs and establishes at one stage in life, under one set of responsibilities, constraints, and technological aids, may be unsatisfactory in another stage in one's life.

The *interpersonal* dimension of privacy refers to the things one does to manage social interaction and information. One's social setting and its physical characteristics provide many options for managing social interaction; physical and social boundaries can be used to control people's access to one another. Information management refers to the patterns of disclosure and nondisclosure that one employs. These two aspects of the interpersonal dimension are interdependent. An individual who has well-developed verbal and nonverbal skills of information management may be willing to enter social and physical settings that do not offer other kinds of devices for controlling interaction and information.

The *control/choice* dimension develops out of one's dimensions of self-ego, culture, and environment. Early in life one has no control over one's privacy, except through hiding. Later one learns to use personal, cultural, and physical resources to exercise subtle control over privacy. Events that would be a threat to one's privacy early in the development of control/choice mechanisms later become so easy to control that they are no longer considered to be a threat to one's sense of privacy, dignity, or self-esteem.

In summary, the need for privacy and the options that individuals develop for controlling privacy depend on age and opportunities to be alone and develop autonomy, on environment--cultural norms, physical settings, and the ways one is permitted to use these settings, roles, and resources in life--and on resources for learning to control interaction and communication. How an adult would go about developing effective and healthy control of privacy in an as-yet-to-be-designed space habitat cannot be known at this time. However, we can presuppose in general what factors might influence the development of a personal definition of privacy and a means for protecting that privacy.

Privacy, Stress and Performance

We have seen that privacy is difficult to define adequately, short of presenting a theory of privacy. However, within the context of the theory presented in the preceeding section, there is heuristic value in the following definitions of privacy:

> Privacy is equated with "control of stimulus input from others, degree of mutual knowledge, and separateness of people from one another" *(9)*,

and

> Privacy is "the selective control over access to the self or to one's group" *(10)*.

These definitions do not tell us what privacies persons need or how they would go about protecting those privacies, but they are heuristic in that they remind us that privacy occurs when the individual can respond to a need for privacy by regulating stimuli so that the need for privacy is satisfied. Viewed in this way, we see that breach of privacy is a form of stress. Persons are said to be placed under stress when they cannot regulate aversive stimuli such as noise, crowding, the flow of tasks they are expected to accomplish, personal criticism, social discrimination, heat, cold, intrusion into one's privacy, and so on. That is, the regulation of privacy according to one's needs is a special case of effective management of stressors. Viewed in this way, the relevance of privacy to performance and health becomes apparent.

Stress is an aversive emotional or physical stimulus to which an individual fails to make a satisfactory adjustment, which in turn causes trauma that contributes to disease. Selye *(11)* and others have shown that when an individual is subjected to repeated aversive stimulation and cannot control or transform it so that it ceases to be experienced as aversive, the individual goes through stages of adaptation known as the *general adaptation syndrome*. First, there is a defensive regulation of the body, through the endocrine system, to systemic injury evoked by stress and worked out by a initial stage of shock. Then there is a stage of growing resistance or adaptation and either the return to health, or, if the adaptation fails, the onset of exhaustion and deterioration of health. The general adaptation syndrome operates such that the individual overcomes the initial stage of shock and is able to function for a while in a nearly normal way, even when the adaptation is not ultimately successful. For example, one who must work under stressful conditions (e.g., where privacy is often invaded, or where work flow or temperature control is unsatisfactory) may respond with initial deterioration of performance but quickly overcome the initial reaction to the trauma. However, unless the adaptation is successful-- unless it enables the individual to control experiences of the stimulus so that it is no longer aversive--other aspects of functioning will be altered. The ability to monitor subtle cues may be reduced. Patience and good manners may wear thin. Muscle tone may change in ways that contribute to exhaustion. Digestion of food may be hindered. Blood pressure may rise.

An unsuccessful adaptation syndrome ultimately results in severe health problems, such as ulceration of the digestive tract or hypertension. Long before then, however, emotional or behavioral problems arise, though the problems that are most obvious may be misleading. For example, failure to perform well may occur not in the stressed individual but in a co-worker who is reacting to the emotional responses of the stressed individual. Or lowered performance may occur in indirect ways, such as through reduced morale or failure to set appropriate priorities, rather than in the form of failure to perform a given task proficiently.

Various kinds of evidence can be offered to support the argument that invasion of privacy is a form of stress and that it leads to attempts at adaptation which, if unsuccessful, are detrimental to health. For example, Wolfe and Golan (12) have shown that invasion of privacy gives rise to feelings of stress and trauma--that persons report experiencing anger when their privacy is invaded, and use terms such as "awful", "hurt", "afraid", and "very upset" to describe their feelings.

Long-term behavioral responses to invasion of privacy vary, depending on the individual; in relatively powerless and emotionally immature persons, withdrawal and acting out are frequent compensatory mechanisms for lack of privacy. While the long-term consequences for health embodied in continual interaction or forced disclosure have not been studied extensively, a few studies exist. For example, D'Atri (13) found significantly higher levels of blood pressure in prisoners who had a cellmate compared with prisoners who had their own cell.

(It should be pointed out that unassailable arguments of either an inductive or a deductive nature cannot be given at this time to show that invasion of privacy functions as a stressor. Inductive arguments are assailable on grounds that it is difficult to separate invasion of privacy from other factors. Hence, any demonstration that invasion of privacy resulted in trauma adaptation and loss of health is assailable on grounds that something else, and not invasion of privacy, resulted in the trauma and illness. Deductive arguments that privacy is a stressor and therefore results in trauma adaptation and possibly illness are also assailable on the grounds of vagueness and subjectivity; privacy may be invaded with any stimulus that the individual *experiences* as an invasion of privacy. Similar problems of definition and of demonstrating causal relationships have challenged those who have done research and formulated theory about stress phenomena *(14)*. This challenge has been met through empirical research that has established patterns in which stimuli that subjects identify as stressors are shown to lead to a predictable set of physiological, emotional, and behavioral responses, which in turn produce either a transformation of the stressor or a decrement in health.)

One of the particularly troublesome aspects of stress is that stressors and the products of stress usually function cumulatively. Thus, for example, a space traveler whose privacy is repeatedly invaded, whose work timelines are impossible to meet, and who creates, as a result of these stressors, much ill will among co-workers will then be stressed by their anger as well, and will have to adapt to all three of these stressors and to any of their additional aversive consequences. While each of the physical and emotional environments may be bearable and surmountable, alone, the cumulative effects of these stressors may be catastrophic, and may take their toll on health and performance before the problems are recognized and resolved.

Problems of stress and stress management will confront space travelers, and it will be critical to the success of space missions and space habitats that these stresses be managed adaptively. Persons who travel in space vehicles and those who eventually live in space settlements will experience unusual and debilitating stresses both from the environment and from work roles. High-level performance will be necessary for survival, particularly in the early stages of space habitation, and the economic costs of poor human performance will be extremely high. The problems of stress and stress management are compounded in two critical ways in space. First, some of the stresses likely to occur will be different from those on earth. Second, societies in space cannot afford to ignore stress as we do on earth; poor workmanship, illness, aggression, violence, and alienation will be harder to absorb into fragile space environments; and there will be no habitable hinterland into which derelicts and refugees from the stress of group living may wander. Thus, it is important that pioneers into space understand the nature of stress, be aware of when and how persons in space environments are stressed, design ways to prevent and reduce distress, and resolve in a rational and principled way the problems of privacy that this implies.

Anticipating Some Conflicts of Interest with Need for Privacy

Just as privacy is relative to many other factors, invasions of privacy are similarly often relative to other factors. Invasions of privacy may be random or accidental events, but more typically they are deeply rooted in the situation. One of the conflicts with the need for privacy that is certain to arise for early space voyagers has to do with the interest of society and of space voyagers in the physical and psychological well-being of voyagers. It is essential, for the reasons previously discussed, that stresses and reactions to stress in space be monitored and documented. Naturally, one highly desirable way to go about this is to send into space individuals who are sensitive, articulate, and open, and who will give an in-depth account of the ways in which they experience and respond to the stresses of their roles in space. There are, of course, a number of reasons why this is an almost unattainable ideal. Those who are willing and eager to go into space may be unwilling or unable to report their experience in a way that would provide scientifically useful data. They may be, by nature, closed and unreflective individuals. They may, by the nature of their existence in space, be unable to focus reflectively on their feelings without jeopardizing their ability to perform their other tasks. They may be unwilling to admit to the level of stress, trauma, and maladaptive functioning they undergo, for fear of being considered by their employers to be unsuited for their work in space and being recalled to earth. They may not feel sufficient rapport with those to whom they report to be able to give a sensitive accounting of their experience. Or they may not trust those to whom they report to control the information adequately, and thus to prevent its being used in embarrassing, inappropriate, or unfair ways. Any of a combination of these and other factors could jeopardize the accuracy of reporting.

Another means for monitoring stress and coping that would be highly attractive to biomedical and behavioral scientists is the use of objective measures of stress and adaptation. Monitoring of cardiovascular functioning, endocrine functioning, muscle tension measures indicative of anger and tension, and body movement by telemetry, and also the videotaping of task and social performance are among the possible means of obtaining independent measures of stress and coping. Unfortunately, any of these measuring procedures could be regarded as an intolerable invasion of privacy by the space travelers on

whom they are employed. The sense that privacy is
invaded is connected with one's sense of *who* is invading
privacy and *how* the information will be used. Two factors
complicate these matters inordinately:

First, the persons doing the monitoring may be
connected in some way with the employment of the individual
whose responses are being monitored. Thus, as in other
aspects of occupational medicine, the information may be
gathered to make employment decisions. Like company
physicians who serve coal miners, asbestos workers, or
athletes, those who monitor the stress and coping of
individuals in space may use that information in order
to make the individuals more productive, to the detriment
of their long-term health. Or, as with physicians who
conduct cardiovascular examinations of school bus drivers,
the information may be used to retire individuals who
do not wish to be retired, on the grounds that they may
not be able to carry out their duties safely. If, indeed,
the data gathered through monitoring of responses might
be used to make employment decisions, or if the individuals
whose behavior is monitored *think* that the data will be
used in these ways, the sense of invasion of privacy may
be overwhelming, and may even lead to attempts to foil
the transmission of such information.

A second factor that complicates the monitoring of
behavioral and physiological responses in space is that
early pioneers into space will fall into the unenviable
category of "famous persons." The public will have an
immense curiosity about every aspect of their physical and
psychological functioning, just as it has shown an
inordinate interest in Hubert Humphrey's cancer, Marilyn
Monroe's gynecological problems, Betty Ford's breast
cancer, and Patty Hearst's physical and emotional condition.
If the space venture is supported through public funds,
the press and others may try to assert that the Freedom
of Information Act places monitored behavioral and biomedical
data in the public domain, open to public scrutiny. All
of this may be done in the interest of satisfying public
curiosity about matters that are normally private.
Information is certain to be interpreted prematurely,
inaccurately, and in ways that violate not only certain
idiosyncratic senses of privacy, but also the constitutional
right to freedom from dissemination by others of false or
embarrassing information about oneself. Any space
voyagers who hold the belief--warranted or not--that
monitored biomedical and behavioral data will become
public will have grave concerns for their privacy.
Indeed, if monitored data are made public, those who

disseminated the information to the public may be sued for violating both their own professional codes of ethics (e.g., Walters, 1974 (15)) and the constitutional rights of others.

Even apart from the use of such data for administrative (personnel) decision making or for dissemination through mass media, there are many problems that need to be confronted: Who owns those data? Was the voluntary, informed consent of the space voyager obtained? What agreement on limited disclosure of information was made? May members of the person's family see the data? May the individuals see their own data? What kinds of information are relevant and important to gather? How long are data kept, and in what form (e.g., microfilm, data tape, etc.)? Are the data kept in an anonymous form or are unique identifiers included in the data? What procedures and penalties exist to deter data-gatherers or others from making unauthorized disclosures?

These kinds of questions, along with the questions of whether data will be used for personnel decision making or be disclosed to the media, need to be explored openly and negotiated in a multilateral fashion before missions are undertaken. If exploration is undertaken while options are still open, any of a variety of solutions can be found.

It is conceivable that space travelers can be selected who have worked out a trust relationship and a set of agreements with their sponsors so that they are willing to have their monitored data used in personnel decision making. Agreements of this kind are possible when reasonable and rational criteria for personnel decision making are spelled out ahead of time and clearly serve the enlightened interests of all concerned, and when everyone is held accountable for abiding by their agreements. It is also conceivable that space voyagers and their sponsors will prove incapable of reaching or honoring an agreement of this kind, and that it will be necessary to create absolute and demonstrable separation of monitored data from administrative or personnel decision making. Between these two extreme solutions, there lie many intermediate possibilities for agreement to *limit* the administrative use of scientific data. For example, there might be an agreement that scientists could disclose monitored data to administrators only if certain specified physiological or psychological abnormalities are observed, and if the abnormalities are at such a level that the welfare of the voyagers may be seriously jeopardized.

Similarly, the public's right or need to know about the well-being of space voyagers should be examined in advance. A policy needs to be worked out with the press concerning the conditions under which information is disclosed. Clear agreements need to be reached about what would constitute legitimate disclosure and what would be irresponsible or fraudulent disclosure. The public's need to "humanize" space voyagers by knowing personal information about identified individuals might be satisfied in ways that are acceptable to and controlled by the individual space voyagers themselves.

Will those who work to explore space have the foresight to explore what privacy may mean in the context of space habitation? Will needs be anticipated, conflicts of interest explored, stresses averted *before* persons and organizations are hopelessly embroiled in the consequences of invasion of privacy? The humanization of space begins with our humanization here on earth--our willingness to anticipate and to discuss openly the way we are likely to feel about ourselves and our privacy in space, and our willingness to work out sensitive and honorable agreements to respect one another.

References and Notes

1. T.M. Cooley, *A Treatise on the Law of Torts*, 2nd edition (Callaghan, Chicago, 1888).

2. A.F. Westin, *Privacy and Freedom* (Atheneum, New York, 1967), p.7.

3. P.A. Dionisopoulos and C.R. Ducat, *The Right to Privacy* (West Publishing Co., St. Paul, Minn., 1976).

4. K. Greenawalt, "Privacy and its Legal Protections," in *Hastings Center Studies 2* (September 1974), pp. 45-68.

5. C. Geertz, cited in Westin, *op. cit.*, p. 348.

6. C. Geertz, cited in Westin, *op. cit.*, p. 16.

7. Roberts and Gregor, cited in I. Altman, "Privacy Regulation: Culturally Universal or Culturally Specific?" *J. Soc. Issues 33* 3 (1977).

8. R.S. Laufer and M. Wolfe, "Privacy as a Concept and a Social Issue: A Multidimensional Developmental Theory," *J. Soc. Issues 33* 2 (1977).

9. Simmel, cited in I. Altman, "Privacy: A Conceptual Analysis," in D.H. Carson (Ed.), *Man-Environment Interactions: Evaluations and Applications*. (Part II, Vol. 6: S.T. Margulis, Vol. Ed.) (Environmental Design Research Association, Washington, D.C., 1974), p. 5.

10. I. Altman, *ibid.*, p.6.

11. H. Selye, *Stress Without Distress* (J.B. Lippincott, Philadelphia, Pa., 1976).

12. M. Wolfe and M.B. Golan, "Privacy and Institutionalization," Paper presented at the meeting of the Environmental Design Research Association, Vancouver, B.C., May 1976.

13. D.A. D'Atri, "Psychophysiological Responses to Crowding," *Environment and Behavior 7* (1975).

14. H. Selye, *op. cit.*

15. LeR. Walters, "Ethical Aspects of Medical Confidentiality," Paper presented at the Sixth Buffalo Conference on Computers in Clinical Medicine, Niagara Falls, N.Y., June 1974.

Richard L. Kline

4. Habitat Requirements, Design and Options

Abstract

This paper matches habitat requirements and design with a spectrum of future manned space missions which could be undertaken in the next two decades. The spectrum starts with manned support systems which can augment shuttle capability, including a Manned Remote Work Station and tended Shuttle Space Platforms, and extends to Space Construction Base applications for Solar Power Satellite construction. Options for low earth orbit and geosynchronous operations are presented, along with techniques for handling emergency and rescue operations.

Discussion

As we enter the 1980s and approach the era of the Shuttle, with its opportunity for routine space operations, we need to examine space habitation provisions which support the growing demand for living and working in space.

Space Habitation Evolution

The three phases of Space Habitation in both low earth orbit (LEO) and geosynchronous orbit (GEO) are shown in Figure 1. Consider the low earth orbit phases first; Shuttle flights are scheduled to LEO starting this year. On these early flights, the Shuttle itself will operate in a *Sortie* mode and will provide the complete life support for space habitation. In the near future it will probably be expedient to provide a Space Platform in LEO which will be the power supply, work bench, and laboratory of the sky, and which will be visited by the Shuttle on a periodic basis. During these early visits to the Space Platform, space habitation provisions will continue to be provided by the Shuttle.

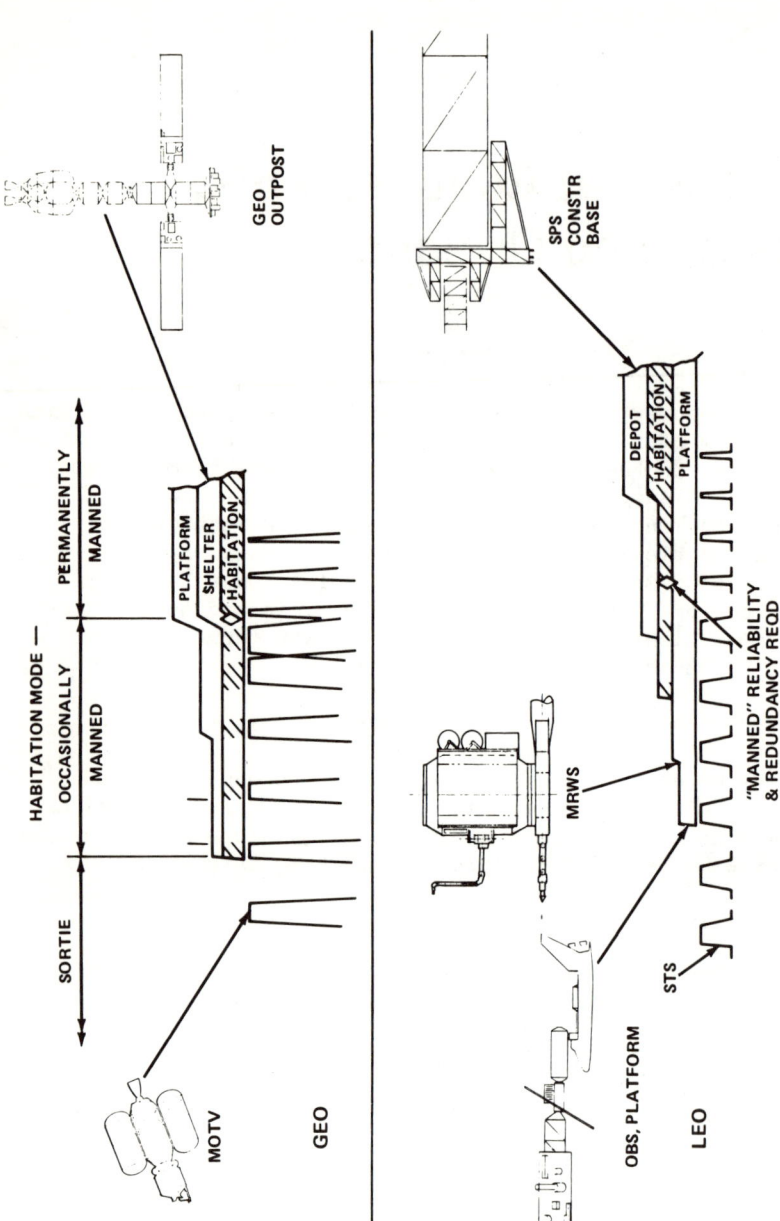

Figure 1. Space habitation evolution.

Habitat Requirements, Design, Options 81

As the activity progresses in LEO, we will move from the Shuttle Sortie Mission Habitation activity to an *Occasionally Manned* phase, as shown in Figure 1, where habitation provisions will be added to the Space Platform so that, when man visits via the Shuttle, extended working and living volume and other habitation provisions will be available to him to enhance his functioning. During this Occasionally Manned habitation mode, the Shuttle will be constantly available as a back-up and safe haven in the event of any need.

With continued activity in LEO, we will reach the third phase, *Permanently Manned*, where now the Shuttle will transport men, equipment, and supplies to the LEO Space Platform with its habitation provisions and then depart, leaving the Space Platform with its crew in space until the next time that the Shuttle arrives to resupply the Space Platform. At this stage, the space habitation provisions must meet all the necessary reliability and safety provisions required for safe support of man in space on a continuous basis.

Manned geosynchronous orbit space habitation evolves through the same three phases, as is also shown in Figure 1. In the first phase, the Sortie Mission, a Manned Orbital Transfer Vehicle (MOTV) fills a role corresponding to that of the Shuttle in LEO. The MOTV consists of a pressurized vehicle, with subsystems for complete manned support, and propulsive capability to transfer it from the Shuttle in LEO up to GEO for use in occasional missions at geosynchronous altitude. The pressurized vehicle with men in it is then returned by the propulsive portion of the MOTV to LEO for rendezvous and docking with the Shuttle, which will then return the crew safely to earth.

In a similar manner as for the LEO scenario, as requirements for GEO missions increase, a Space Platform with special equipment and habitation provisions will be added to extend the effectiveness of MOTV operations. At that point, and in direct parallel with LEO operations, manning will consist of Occasionally Manned occupation of the Space Platform with its life support equipment, while at the same time the MOTV affords a safe haven and a vehicle for emergency return to LEO, if required.

Eventually, the time will once again be reached when Permanently Manned occupation in GEO is required. The habitation provisions attached to the geosynchronous Space Platform will then have to be reliable as a guarantee of safety for the Space Platform inhabitants when the MOTV is

MISSION	ORBIT		ACTIVITY	CREW			HABITATION MODE
	LEO	GEO		SIZE	SKILLS		
• SATELLITE SERVICING	▨	▨	REPAIR UPDATE	3/4	PILOT/TECH		SORTIE
• OBSERV PLATFORM	▨	☐	REPAIR UPDATE CALIBRATE OPERATION	3/4	PILOT/SCIEN/ TECH		SORTIE/ OCCASIONAL/ PERMANENT
• SPACE PROCESSING	☐		DEVELOP PILOT PLANT PRODUCTION	2/3 4/5 4/5	PILOT/SCIEN/ TECHN SCIEN/TECH PILOT/TECH		SORTIE/ OCCASIONAL PERMANENT OCCASIONAL
• DEPOT OPS	☐		OTV TURN-AROUND CARGO TRANSFER	6/12	TECH		PERMANENT
• CONSTRUCTION – EARLY & INTER	▨		FABRICATE ASSEMBLE CHECKOUT	5/10	PILOT/CONSTR/ TECH		OCCASIONAL/ PERMANENT
– SOLAR POWER SATELLITE	☐	▨	FABRICATE ASSEMBLE CHECKOUT	100/500	CONSTR./TECH/ SCIEN/HOUSEKEEP		PERMANENT

Figure 2. Man's functions in space.

no longer present. Shelter provisions will be necessary for protection during certain emergency conditions, which will be discussed later.

These, then, are the three modes of habitation in LEO and GEO which, in time, can be employed. In view of the very much greater cost of providing increasingly autonomous space habitation systems, the questions are: What mode should be provided, and when should it be provided? To answer these questions, we need to consider man's functions in space.

Man's Function in Space and System Concepts

Figure 2 collects the functions that man can perform in space into six mission categories, specifies the orbit, crew size and activity, and characterizes the habitation mode. The missions start with Satellite Servicing, which, as is indicated in the figure, is carried out using the Sortie habitation mode by a small-size crew in LEO or GEO, as appropriate. At the far end of this spectrum is Solar Power Satellite (SPS) fabrication, assembly, and checkout, which requires a Permanently Manned space habitation mode for crews to perform it. This operation, involving several hundred people in space, represents a much more advanced capability in both LEO and GEO. Examples of concepts that meet these mission requirements are illustrated in the following figures and discussion.

Satellite Servicing, an early Sortie mode of operation, is illustrated in Figure 3. The Shuttle is employed for manned life support in low earth orbit, and a mobile worksta- tion, or "Open Cherry Picker," is shown attached to the end of the Shuttle Remote Manipulator System *(1)*. The spaceworker performs the Satellite Servicing mission by replacing defective modules on the satellite with fresh equipment. The Open Cherry Picker workstation requires a vice-like function shown at the bottom of the platform, where a stabilizing arm has been attached to the satellite to remove differential motion. Lights are strategically located on the Open Cherry Picker to provide both general and point illumination during the activity. The small display panel for controlling the position of the Open Cherry Picker swings away when it is not needed so that the spaceworker's view and reach are not obstructed. A tool and parts box behind the astronaut reduces the need for frequent returns to the interior of the Shuttle.

Space Observation and Space Processing missions in LEO give rise to the need for Space Platforms to extend the

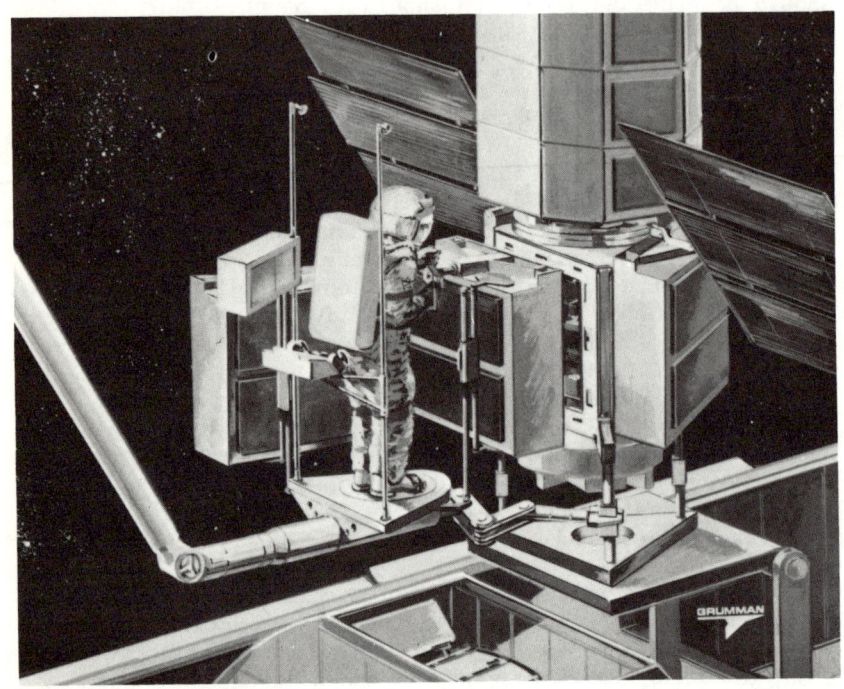

Figure 3. Open cherry picker.

Habitat Requirements, Design, Options 85

useful time in space, provide added electrical power and heat rejection, and lower costs for individual industrial activities. Figure 4 illustrates one Space Platform concept (2) that meets the needs listed in Figure 2 and has the necessary provisions for Occasionally Manned operations. The Habitation Module consists of a large Spacelab Module which extends the living and working area. The Shuttle docks to this module at each visit so that there is "shirt-sleeve" transfer from it to the Habitation Module. The environmental control cycle is still not closed; oxygen and water are supplied and resupplied by the Shuttle. This concept can transition into Permanently Manned space habitation capability with additional modules and life support systems.

An example of a Habitation Module for LEO construction missions is the Closed Cabin Cherry Picker, illustrated in Figure 5. The Closed Cherry Picker provides a pressurized working volume with dexterous manipulators so that space-workers can perform their space construction functions without wearing cumbersome and tiring space suits. The Closed Cabin Cherry Picker is itself fastened to an advanced Remote Manipulator System, or crane, so that it can be maneuvered around the worksite and can return rapidly to the LEO Space Platform and the Shuttle, which provides the basic habitation support.

The same concept could be employed in GEO operations for Satellite Servicing. Because of the higher radiation levels the Cherry Picker must be of the closed variety, with sufficient shielding to keep the radiation dosages at safe levels.

GEO Satellite Service Missions, as defined in Figure 2, can be accomplished initially using an MOTV as pictured in Figure 6. The propulsive stages and Habitation Module are brought to LEO by the Shuttle and then assembled prior to transportation to GEO.

The interior of the Cabin Module is shown in Figure 7. This arrangement features a pressure vessel, sized for four crewmen and divided into three zones: mission task, crew support and equipment installation. The habitation area has four individual crew quarters, together with waste management and personal hygiene compartments and a food preparation area. Dining equipment is movable for access to an emergency escape hatch. A large central area is used for rest, recreation, and physical exercises.

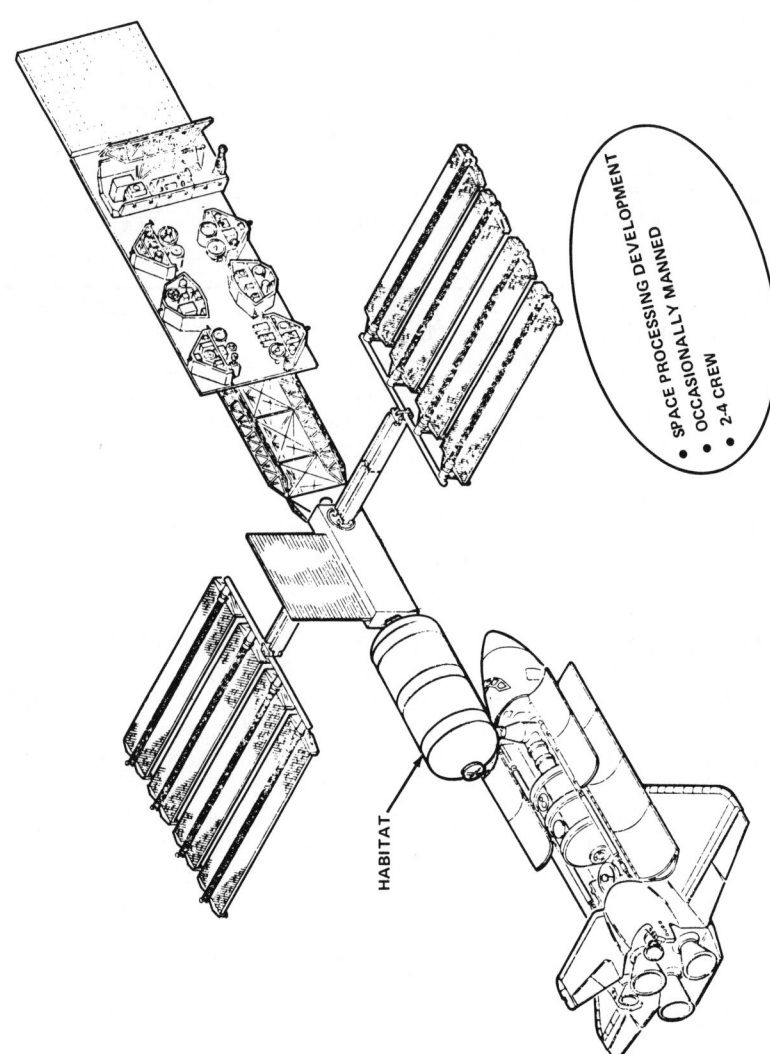

Figure 4. LEO observation platform.

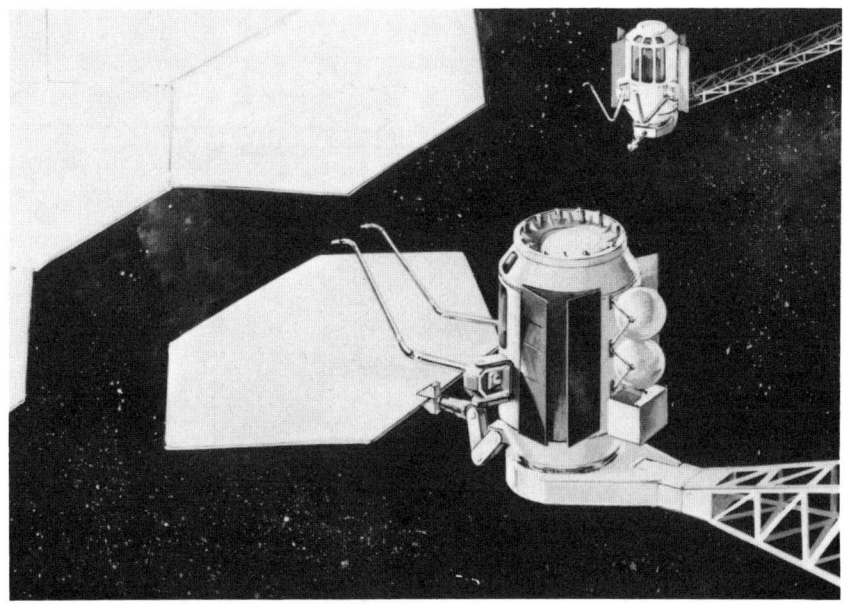

Figure 5. Closed cabin cherry picker.

Figure 6. Manned orbit transfer vehicle (MOTV).

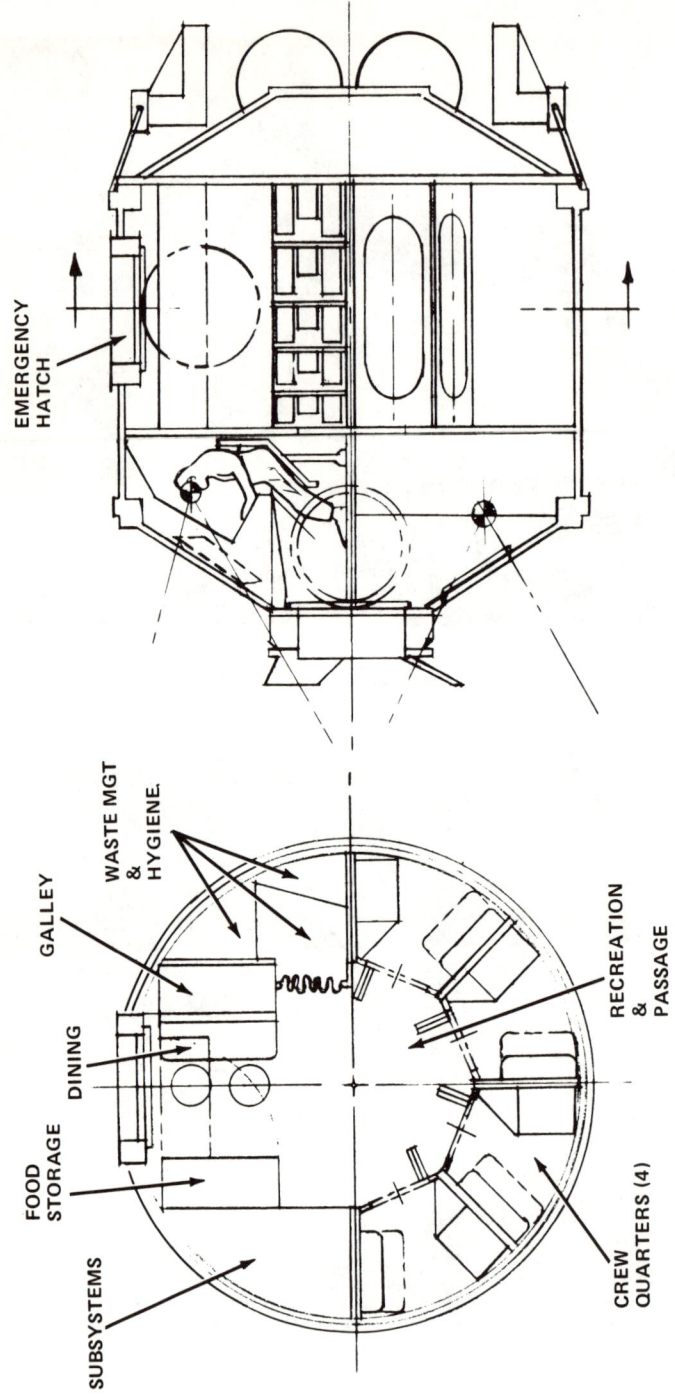

Figure 7. Cabin module for manned orbit transfer vehicle.

Habitat Requirements, Design, Options 89

The Solar Power Satellite Construction Mission requires very extensive manned operations for the production units, which will generate 5000 to 10,000 MW of electrical power for earth use. Space Construction Bases, incorporating the technology developed during the much more modest construction activity illustrated in Figure 5, will now require habitation facilities similar to those of a substantial naval ship. One Space Construction Base concept and its Control Center are shown in Figure 8.

In similar fashion, the features of the Space Platform and habitation concept suited to the other mission modes of Figure 2, and variants on them, can be developed. In each case the concept must meet general habitability requirements such as life support, pressurization, docking/airlock, lighting, first aid, etc.

Special Habitation Requirements

Figure 9 is a partial listing of additional Habitation Requirements which must be satisfied, and which provide a contrast between LEO and GEO operations.

The first of these particular requirements is to provide that protection necessary to keep high-energy particle radiation dosages within safe levels for the crew. While there is some uncertainty about tolerable dose rates, a total exposure to the skin of 105 rem over a 90-day period has been used in many space design studies.

That value is not applicable at GEO, where the radiation environment is dominated by electrons. Shielding must be provided both against trapped electrons and against *bremsstrahlung*, which result from the decay of electrons to photons when they meet shielding. A low-weight protection system is shown in Figure 9, in which the shielding is a composite, or sandwich, of materials. The outer shielding is a material of low atomic number--in this case, epoxy--to inhibit the production of photons. The pressure shell is aluminum and, since it must be 8 mm in total thickness, it is shown as two thicknesses to provide redundant pressure vessels. The inner shielding material has a high atomic number--in this case, tantalum, which is very effective at blocking soft *bremsstrahlung*. Although this is an efficient system and is relatively low in mass, it is three times heavier than that necessary only to hold cabin pressure.

This puts a high premium on Cabin Volume in GEO because now the Cabin--which must be provided for habitation--weighs three times as much as a LEO cabin and has a cost to orbit

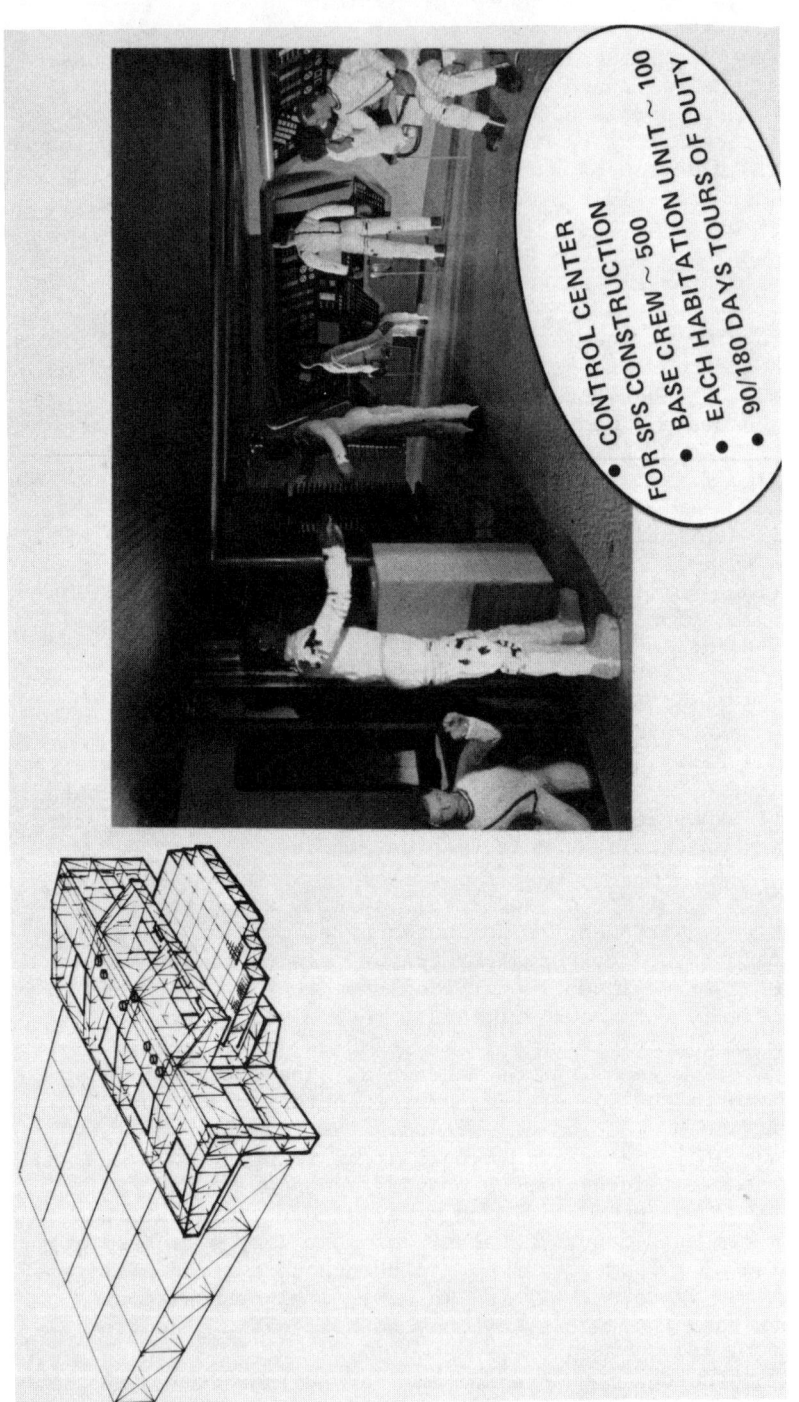

Figure 8. Permanent large-scale space habitation.

	LEO	GEO

• RADIATION PROTECTION
  – CABIN
 	
  – SOLAR FLARE SHELTER
	ALUM SHELL (4mm)	EPOXY TILES (5mm) ~ OUTSIDE
		ALUM SHELL (8mm TOTAL)
		TANTALUM (0.5mm) ~ INSIDE
		
		SHELL = 150mm ALUM

• VOLUME PER MAN
  – CABIN
  – SOLAR FLARE SHELTER
	8–10 m^2	8–10 m^3
		0.6 m^3

• RESCUE – MAX DELAY	5 DAYS	30 DAYS

• MAINTENANCE
  – BY EVA
  – COST PER MAN HOUR
  – TRANSPORT COST OF RECYCLED COMPONENTS
	ACCEPTABLE
	$2,500
	$1,000–1,800/kg
		VERY LIMITED
		$9,000
		$10,000/kg

Figure 9. Habitation requirements.

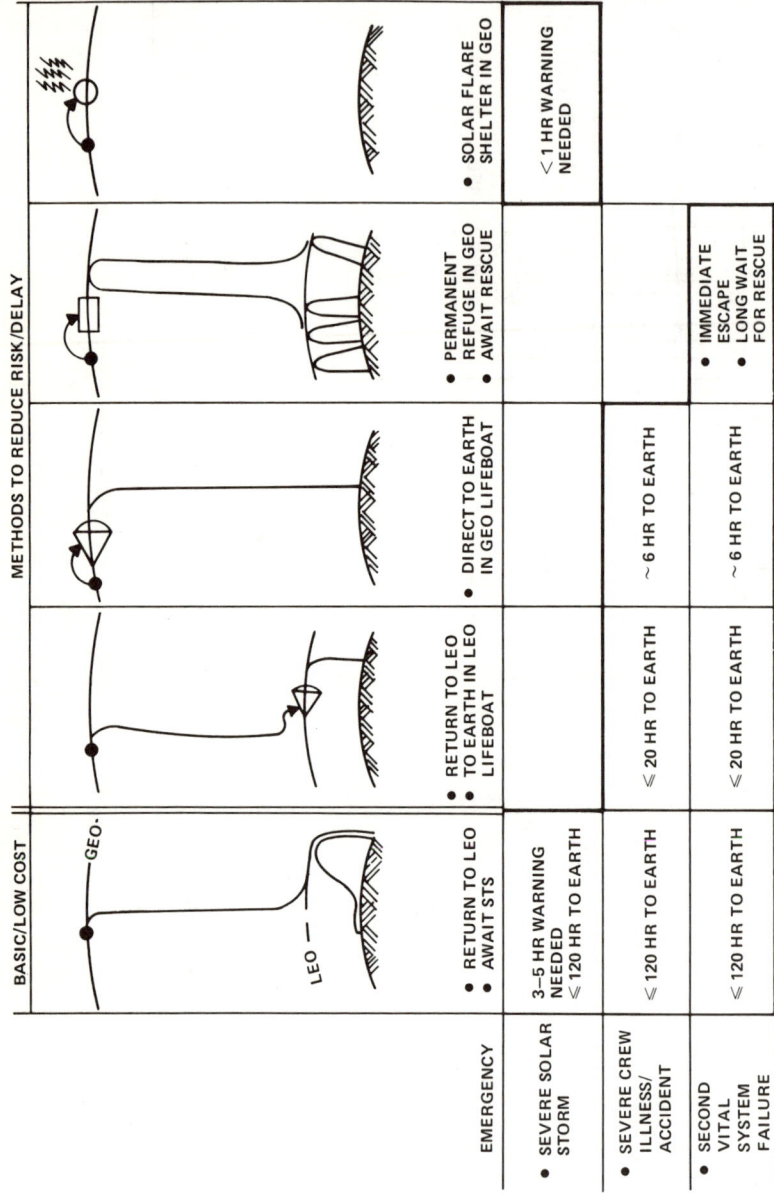

Figure 10. Responses to severe emergency on GEO sortie mission.

that is five times as much for each kilogram transported. Strategic placement of the equipment surrounding the pressurized cabin can reduce this penalty by helping to absorb radiation, thereby reducing cabin weight, but the impact of weight remains high for GEO operations.

Turning to the Maintenance issue, this is also quite different in GEO from LEO. Simple repairs which could be performed by a crewman in EVA suit in LEO are greatly limited in GEO because of the higher-radiation environment. The cost/man-hour in GEO is more than three times that in LEO, which, when combined with an equipment transportation cost to GEO more than five times that in LEO, greatly changes the Space Platform maintenance philosophy. Large cost premiums can now be paid for high-reliability equipment and components for GEO operation, even in the early phases. Built-in redundancy, rather than in-flight repair techniques, is indicated.

The Rescue issue is quite different between LEO and GEO. In GEO, solar flares are a threat which can be met by providing a storm shelter within the Habitation System. Since a wall thickness the equivalent of 150 mm of aluminum is required, the shelter size must be kept as small as possible to minimize the weight penalty (about 2,200 kg/man for a four-man shelter). An alternative to providing a shelter would be a plan to evacuate the GEO Space Platform and return to earth or to a safer altitude in LEO in such an event.

These latter responses to emergency require more time to carry out. In the case of a solar flare, 3-5 hours' warning time is needed to perform a successful escape to LEO. The shelter in GEO, by contrast, affords much greater safety in that only as much warning time as is necessary for the crew to enter the shelter is required.

There are other reasons for rescue than a severe solar storm, including severe crew illness, accident, or a second failure in a vital Fail Operational/Fail Safe System that pertains to crew safety. Rescue is far more difficult in GEO than in LEO. Rescue time might be 30 days for GEO, as against five days for LEO, because depot operations in LEO to assemble an MOTV vehicle would take several Shuttle launches. Several techniques for responding to a severe GEO emergency in an MOTV vehicle like the one shown in Figure 6 are outlined in Figure 10.

The basic and lowest-cost way of handling the emergency is to return the MOTV with its crew to LEO, wait for a

Shuttle to be sent up, rendezvous and dock with it, and return to earth. This method is applicable to all three emergency categories in Figure 10. Depending on the phasing of the orbit, however, it may take up to 120 hours elapsed time to complete.

A return-to-earth time of up to 120 hours may not be acceptable in the case of severe crew illness or system failure, and this is where the other methods shown in Figure 10 fit in. One "lifeboat" technique consists of returning to LEO, rendezvousing with a re-entry capsule already in LEO, and returning to earth in it. Additional equipment must be developed, but rescue time in that scenario would be reduced to less than 20 hours. A second method is direct re-entry in a GEO lifeboat, which would reduce rescue time to about six hours.

Which method or combination of rescue methods is finally selected will depend on an evaluation of risk definitions for the particular vehicle and the operations being accomplished.

Conclusions and Recommendations

Several observations and conclusions may be reached regarding space habitat applications and designs:

- Techniques are available for supporting living and working in space, many of which have been or will be demonstrated by the Apollo and Skylab programs. We can be confident, therefore, that we can provide habitat features to support near term expanded manned space operations via the Shuttle.

- The classifications of potential manned missions point to additional new roles for man, including blue-collar space work in large-structure assembly and fabrication.

- Habitat concepts vary substantially, depending upon the mission requirements. Indeed, specific definition and time scheduling of the mission activities should be the driver for the overall habitat approach.

- Manned GEO operations differ from LEO operations in several significant ways that impact the habitat concept. We must be alert to these differences and establish effective approaches for meeting the new requirements.

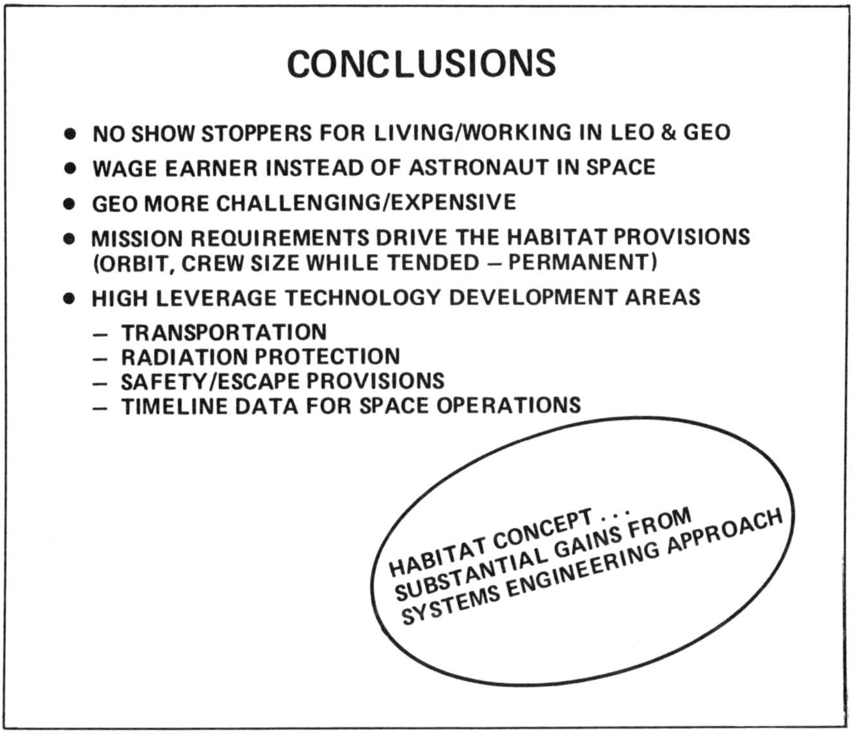

CONCLUSIONS

- NO SHOW STOPPERS FOR LIVING/WORKING IN LEO & GEO
- WAGE EARNER INSTEAD OF ASTRONAUT IN SPACE
- GEO MORE CHALLENGING/EXPENSIVE
- MISSION REQUIREMENTS DRIVE THE HABITAT PROVISIONS (ORBIT, CREW SIZE WHILE TENDED — PERMANENT)
- HIGH LEVERAGE TECHNOLOGY DEVELOPMENT AREAS
  - TRANSPORTATION
  - RADIATION PROTECTION
  - SAFETY/ESCAPE PROVISIONS
  - TIMELINE DATA FOR SPACE OPERATIONS

HABITAT CONCEPT... SUBSTANTIAL GAINS FROM SYSTEMS ENGINEERING APPROACH

Figure 11. Conclusions.

Although there is no apparent "show-stopper" involved in extending man's habitation in space to meet growing requirements, there is still much to accomplish if we are to have the technology ready in the 1980s. Figure 11 lists some of the high-leverage technology development areas which I believe need emphasis, in addition to work which is currently under way.

At several points, this paper has illustrated the fact that the habitat concept has major impacts on many elements of the overall manned space operation. Its development therefore requires a broad systems engineering approach, to ensure that all real requirements will be met in an efficient manner.

Sound manned space habitat design and technology is a vital key to helping us capitalize on the Shuttle and evolve toward effective utilization of space. Added systems engineering and technological development activity of the kind discussed in this paper should be planned and carried out to assure that we can expand man's space operation as the requirements demand.

References and Notes

1. C.A. Nathan, *Manned Remote Work Station Study*, First Interim Report (No. NSS-MR-RP006, Grumman Aerospace Corporation, Bethpage, N.Y., 1978).

2. J. Mockovciak, *Large Space Structures Demonstration Study*, Mid-term Report (No. NSS-LS-RP016-1, Grumman Aerospace Corporation, Bethpage, N.Y., 1978).

_____*Allen J. Louviere, John T. Jackson*

5. Man-Machine Design for Spaceflight

Introduction

The environment of weightlessness is the single significant difference a designer must consider for human beings and machines in space vehicles. This characteristic fascinates the journalist, intrigues the space enthusiast, and provides a new challenge to the man-machine designer. Weightlessness literally lifts the burden of gravity from mankind's shoulders. This, in itself, may be as near as man will come to soaring into unpowered free flight. Even with these newly found capabilities, effective designs for man to live and work in (and with) require diligent attention to the subtleties of the changes that are peculiar to the weightless environment.

Initial designs illustrating the differences that appear in weightlessness were tested during the Skylab program. Never before had men experienced anything like Skylab, from the standpoint of time spent in space as well as from the standpoint of the large size of the space station, both of which imposed special sets of circumstances on the astronauts. Previous space missions had amounted to making relatively quick trips through space in vehicles no larger than a car. The Skylab astronauts were the first successfully to take up residence in the weightless environment for a long stay. The nine Skylab astronauts spent a total of a hundred and seventy-one days in space, circled the world two thousand four hundred and seventy-six times, and traveled seventy million, five hundred thousand miles--farther than any men had ever traveled before, and about 10 times as far as the total distance traveled by all the Apollo crews who went to the moon. Men had never before experienced anything like Skylab.

For at least three and one-half billion years—ever since the first cell formed—man's body, both as a species and as an individual, has been shaped by continuous exposure to gravity, so that a large portion of it is by now dedicated to more or less continuously opposing gravitational forces. In space the effects of gravity vanish, and, with them, most of the reasons men and other creatures of the earth are the way they are. Therefore, one could confidently predict that placing the human body in weightlessness would produce changes in size, shape, and composition. From the earliest days of spaceflight many of these changes and their effects were described by astronauts.

Anthropometric Changes in Weightlessness

In the weightless environment of space, these rather profound changes in anthropometry take place rapidly. There are immediate anthropometric changes in height. Without gravity pressing down on the spine, the spine tends to unload and the astronauts grow approximately two inches during the first day or two of weightlessness. They retain this increase throughout the mission until reexposed to earth's gravity, when the process reverses itself. Virtually all this increase in height, it appears, is caused by lengthening of the spinal column, and the change is limited to the trunk and neck. Although it was not recognized at the time, this height growth impacted space suit design as far back as Apollo 16. On that mission it was almost impossible for one of the astronauts to get his space suit on and zipped up prior to going out onto the lunar surface. On earth this particular astronaut had always liked his suit to fit snugly. In the 1/6 gravity of the moon his height growth was just enough to make it almost impossible to get him into his suit and get it zipped up. The designers were not aware of this height "growth" phenomenon and, consequently, did not realize what to attribute the difficulty to.

Another important anthropometric change that occurs during spaceflight is the loss of body weight. Virtually every astronaut and cosmonaut has lost weight during spaceflight. Weight loss has been an apparently constant side effect of spaceflight. It now appears that most of the weight losses in flight were caused by inadequate diet. Energy costs on Skylab were surprisingly high and reflect the pace of crew activity. One of the Skylab astronauts wondered how it could be that in space, where they seemed to exert themselves less than on earth, they consumed the same amount of calories and still lost weight. The Apollo and Gemini astronauts had also lost weight, but they had eaten less than they did on earth. Part of the answer may have been that the Skylab astronauts were burning up more calories than they suspected in just getting about inside the Skylab workshop, which was 33 times as large as the crew compartment of Apollo.

Understanding the quantity of food an astronaut needs and then providing the required amount in such a way that the astronaut actually consumes it in a systematic manner is vital, not only for the maintenance of good health and high energy levels but for the conduct of the mission itself. On future flights of a year or more into deep space, for example, even a slight miscalculation of daily food requirements could mean excessive weight and volume penalties to the spacecraft.

Concepts of Percentiles in Design

A major anthropometric concern in the design and evaluation of men and machines for spaceflight is the statistical description of all those persons who may, throughout the life of a product's usefulness, be involved in its operation, maintenance, consumption, or other anthropometric relationship with the product. This group of persons may be defined as the total user population. Early in the definition phase of design, anthropometric limits are set in much the same way that tolerances and fit are selected in the design of machine parts. Too tight a tolerance excludes many persons in a design population and places the emphasis on selecting operators within a specified size range. Conversely, too loose a tolerance will include such a large range of persons as to add complexity for adjustments, which increases costs to prohibitive levels.

Shuttle design limits and selection standards for astronauts are specified according to a range of anthropometric data converted into percentiles. Percentiles are the single most effective way of presenting anthropometric data. It is a common practice to design aerospace vehicles to accommodate operators within a range such as the 5th to 95th percentiles. For example, the Shuttle anthropometric design range was specified to include the 5th to 95th percentile males and females. In effect, this says that the smallest 5 percent of the female population and the largest 5 percent of the male population will be excluded from Shuttle.

Percentile ranges not only help reduce data to a manageable proportion but also help strike the proper balance between special and standard design for workstations. Such a range defines the theoretical design limits for the

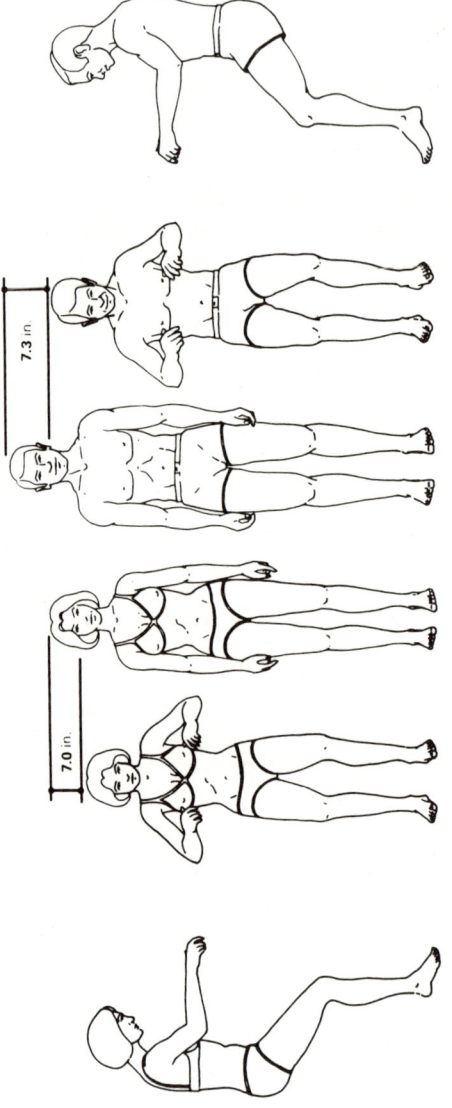

Figure 1. Two center figures show bodies at Earth positions, while four outer figures show neutral body positions of space.

persons actually to be "installed," in effect establishing a
design criteria population. However, it must be remembered
that the concept can yield misleading interpretations and
cause large errors in design when improperly applied.

While the percentile concept of design limits is widely
used, it has acquired some unfortunate interpretations. For
example, the 5th and 95th percentile values from the design
population have become accepted as the only operating design
values for accommodation of the population, and the dimensional values have become formulated as the 5th and 95th
percentile body form, head form, etc. Designers have then
worked to design to the sizes or shapes of variance in these
forms with the idea that by so doing they would accommodate
in their design all the possible combinations of body size
and shape that fall within these limits. The foregoing is
not meant to imply that the concept of 5th and 95th percentile values is worthless as a designer tool, but to point
out that the concept has obvious limitations.

Posture

Skylab studies and film analyses indicate that an
astronaut's neutral body posture differs from that assumed
in a gravity environment. On earth, where everything has
weight, skeletons are needed for rigidity and leverage.
Muscles are needed for sustained motion. The circulatory
system is needed to pump blood against gravity. A considerable part of our daily metabolic output is expended by the
postural, or antigravity, muscles. Man's musculoskeletal
and neurological systems are dedicated to maintaining
upright posture under the force of gravity. In weightlessness, the body has a totally new situation. Released from
the force of gravity, man assumes a neutral position
(Figure 1).

Not only are the large antigravity muscles and associated servoloops unopposed by gravity but various positions,
such as sitting and bending, which are dependent on gravity
for loading are inappropriate. In weightlessness the
relaxed, unrestrained human body automatically assumes and
indefinitely maintains a semicrouched posture. Characteristics of this weightless posture include plantar flexion of
the feet and flexion of hips and knees with slight abduction
of the legs. The thoraco-lumbar spine is straightened or
even slightly flexed forward. The limbs and head move to
the midpoint of their mobility range. The shoulders tend to
move upward with the arms and elbows elevated. Although the
cervical spine (neck) is straightened, it is also angled
forward, a position forcing the head down and thus lowering

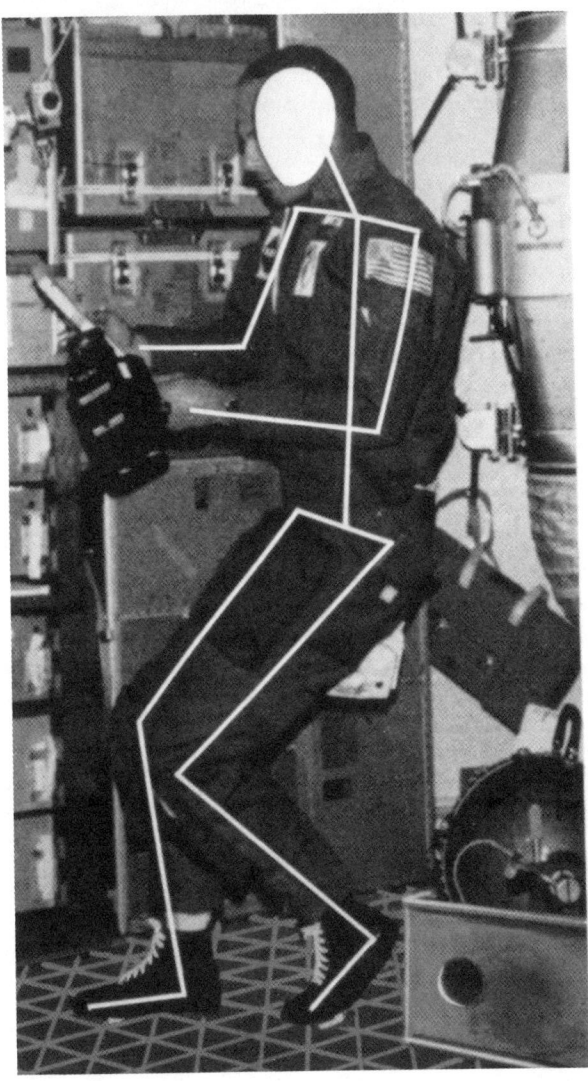

Figure 2. Geometry of the weightless body.

the normal line of sight. Such changes, plus a constant body-fluid redistribution, will tend to shift the center of mass of the whole body headward. The result is that overall shape and posture have been significantly changed by the weightless environment. The body assumes a semicrouched position where limbs and head move to the midpoint of their mobility range. Figure 2 illustrates the weightless body geometry.

New design approaches based on these data would change the entire man-machine philosophy. The natural body posture places important functions in different positions, relative to that of an earth-like environment. The eye point and limb positions are of particular interest with respect to displays and controls. For example, the standing height of a 95th-percentile crewman is 74.3 inches in an earth-like environment, while in a weightless environment the standing height for the natural body posture is only 67.0 inches. The 7.3-inch difference is substantial and impacts both the crewstation design and arrangement and the crew's performance. Similarly, the difference for a 5th-percentile crewman is 7.0 inches (see Figure 1).

The implications and applications of natural body posture must be taken into account if efficient man-machine design for spaceflight is to result. Each element of the design must be examined in the light of weightless considerations. Every feature must be examined to see if gravity influenced the design. For example, where there is no gravity, chairs, couches, beds, and other devices to reduce fatigue are useless for that purpose. On Skylab, the seat at the Apollo telescope console was little used by the first crew and discarded entirely by the second and third crews.

The natural body posture associated with weightlessness must be accommodated if fatigue and discomfort are to be avoided. Having to maintain certain positions may in weightlessness produce much more stress than doing so on earth would produce, since muscles might be called on to supply forces that are normally supplied by gravity. Stooping and bending are examples of positions that always cause abdominal fatigue in weightlessness. The heights and angles of weightless posture must be accommodated.

In weightless natural body posture, the stance is not vertical since the hip/knee flexion displaces the torso backward, away from the footprint. Height is now located at a point between sitting and standing; therefore work surfaces must be higher than those designed for normal sitting tasks. Elevation of the shoulder girdle and arm

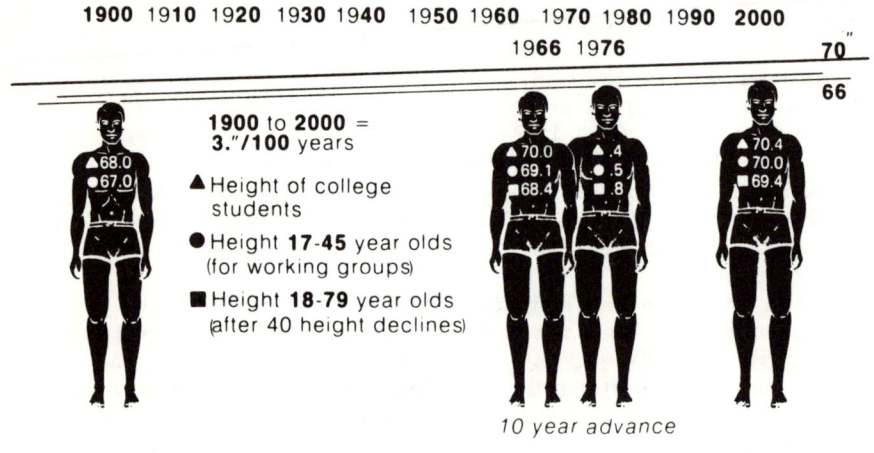

Figure 3. Trends in individual growth.

flexion also makes elevation of the work surfaces desirable. Although in weightlessness the head is angled forward and down, a positioning which depresses the line of sight, eye-to-work level may remain practically the same.

The natural body posture, then, is the basic posture that should be used in establishing workspace layouts and design. A major difficulty in using existing anthropometric data for spaceflight applications is that the data are for the most part static data collected in the gravitational environment of earth. Extrapolating these data to the weightlessness of space is difficult at best. A small amount of inflight anthropometric data was gathered during Skylab. This effort should be expanded in future flight.

Trends

Aside from the newly determined body posture, another factor providing additional requirements to the designer is the general increase in overall size in the United States population. The graphic illustration of Figure 3 projects the average size of the astronaut population of 1985.

Figure 3 clearly indicates a growth trend. The fact that the size of the adult human body is thought to be increasing probably comes as no surprise. The man on the street will tell you that people are getting taller; older members of a community will recall "when people were smaller." Evidence shows that today's children reach peak height earlier in adolescence and each decade sees them reach puberty four to five months earlier than in the preceding one.

The growth trends that occur from generation to generation are usually referred to by anthropologists as secular changes. Whether these trends are the effect of better nutrition, improved health care, or some biological selection process has not been determined and is, in any case, of no practical significance to design engineers--who need to know "how much," rather than "why." The lengthy lead time required for the conception, design, and production of spacecraft is such that the persons who will eventually use them are, for the most part, only children when the design specifications are fixed. It is of more than casual interest, therefore, to anticipate the dimensions of physical size and body proportions which will exist at a given time in the future.

The problems of designing for a highly variable population are immense, but not insurmountable. The solution lies

in a thorough acquaintance with the problem. Equipment and
system designers must arrive at a design solution which will
accommodate the irregularities of size, shape, and mobility
of potential users. It is of value, therefore, to have as
detailed a quantification of body size variability of the
user population as possible.

The differences between the sexes represent a major
source of variability, with the female having, in general, a
smaller overall body size, less strength, and less rugged
features than the male.

Size Differences Between Sexes

Size difference between the sexes is a primary source
of variability. Since NASA now has women astronauts, that
difference in size is a matter of considerable importance to
designers. Areas of design which a few years ago would not
have required the consideration of female size and strength
now do so.

In general, women tend to be smaller than men. In
addition, the sex-based tendency for females to deposit
subcutaneous fat makes women more rounded. In attempting to
assess quantitatively the size differences between men and
women, care must be exercised in selecting data for
comparison.

If male-female differences in the mean value for most
body dimensions average only about 8%, what is the significance for designers? The answer to this question may be
approached in several ways. One method is to examine the
range of size differences, especially for dimensions commonly
used in design, between a small female (5th percentile) and
a large male (95th percentile). This is a realistic approach
for the Shuttle, where persons representative of each
extreme (in one or more dimensions) may be required to use
or operate the same item or function in the same workspace.
From a design viewpoint this size difference between the
sexes means that systems or equipment specifications based
on the anthropometry of male fliers, for example, would have
to be modified to accommodate the body size differences of
female users.

Variation Between Whites and Blacks

Another important anthropometric consideration in
design is related to the variation in body size between
whites and blacks. This variability is of considerable
interest to designers because whites and blacks make up the

largest segment of the American population. Comparisons
between whites and blacks have been made using military
anthropometric data. Height and weight values for both
whites and blacks were very similar. Despite this, there
are significant differences in the mean values for about
three quarters of the measurements. Blacks' legs, arms,
hands, and feet are, on the average, longer than those of
whites; whites show longer measurements of the torso.
Blacks also tend to have longer heads, wider faces, and less
body fat. While individual values for whites and blacks
overlap to a large extent, the body-size differences are of
sufficient magnitude to warrant consideration in the design
of systems and equipment that are to be used by both whites
and blacks.

Variation Among Nationalities

Another major source of variability is found among
persons of different nationalities. While all living people
belong to a single biological species, that species, like
other life forms, is not geographically uniform. It is
differentiated into a number of local variants. These
variants frequently differ in a number of morphological
traits such as skin color and body size and proportions,
with a particular trait often highly characteristic for a
single variety.

It is not necessary here to probe for the reasons
behind these morphological differences, but only to acknow-
ledge their existence and attempt to deal with them in terms
of sizing and design requirements. This variability is of
some importance in this discussion because of the many
ethnic and racial groups that constitute the American population and thus the potential design population in the NASA
space program.

Variation Among International Crewmen

Consideration must be given by the designer to the
marked variations in anthropometric sizes among the international crewmen who are scheduled for some of the Shuttle-
Spacelab Missions. Figure 4 illustrates the range in height
that the designer is confronted with when designing for
users as diverse as the Japanese female and the large
American male. This range of 16.7 inches is an extremely
large one and can present some very difficult--if not
impossible--design problems. From a design viewpoint, such
a range indicates that systems or equipment specifications
based on the anthropometric criterion of 5th to 95th

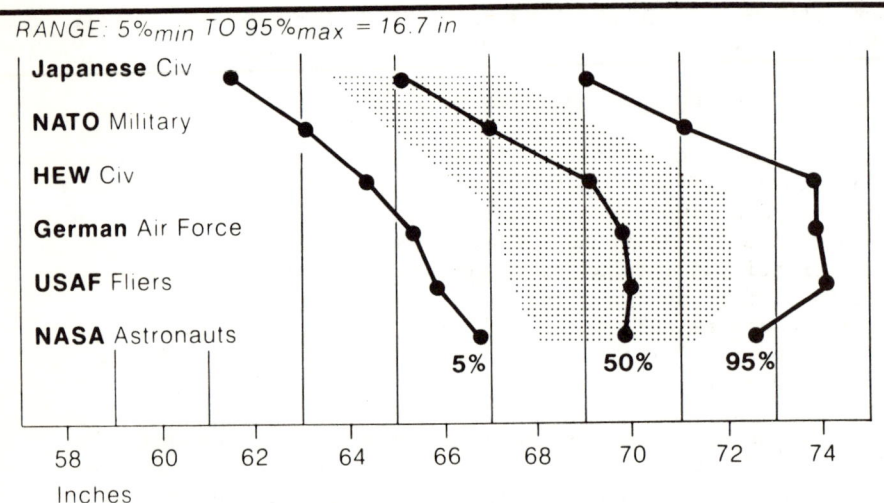

Figure 4. Crew size variations: American, foreign and international.

percentile U.S. male, for example, would have to be modified if they are to accommodate this body size difference.

The total effects of variations in anthropometry as a result of the weightless environment, the increase in the size of the individual, and sexual, racial, and national differences have defined a new design criterion for certain aspects of crew accommodations. The summary effect is that a broader range of crew sizes must be considered and accommodated. Also, these crew provisions will be larger in size. Other pertinent design variations may surface once a data base of flight experience has been accrued.

A Basic Element of Shuttle Workstation Design

For Shuttle operations only foot restraints have been baselined. The concept is shown in Figure 5. Consequently, Shuttle workstations are designed for crewmen in an upright/ standing position. Chairs, seats, etc., provided are only for certain portions of the missions, such as launch, reentry, and orbital maneuvers. The combination of foot restraints and the neutral body posture provides the basic elements that crew workstations were configured for. The Shuttle program will form the reference base for future programs.

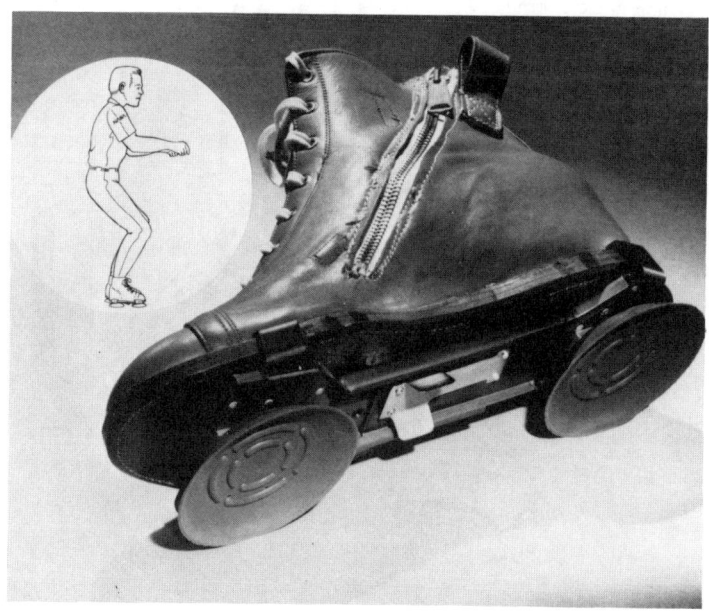

Figure 5. Foot restraint.

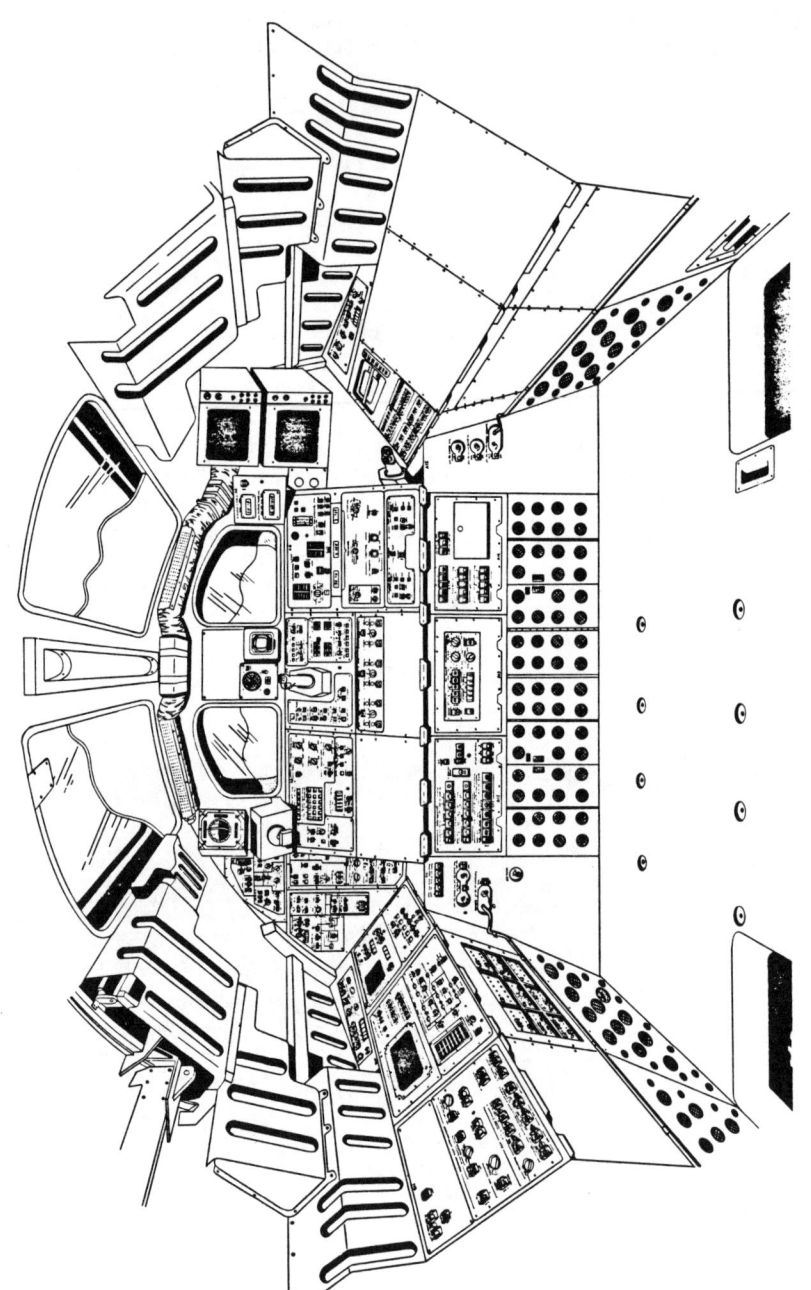

Figure 6. RMS Control Station.

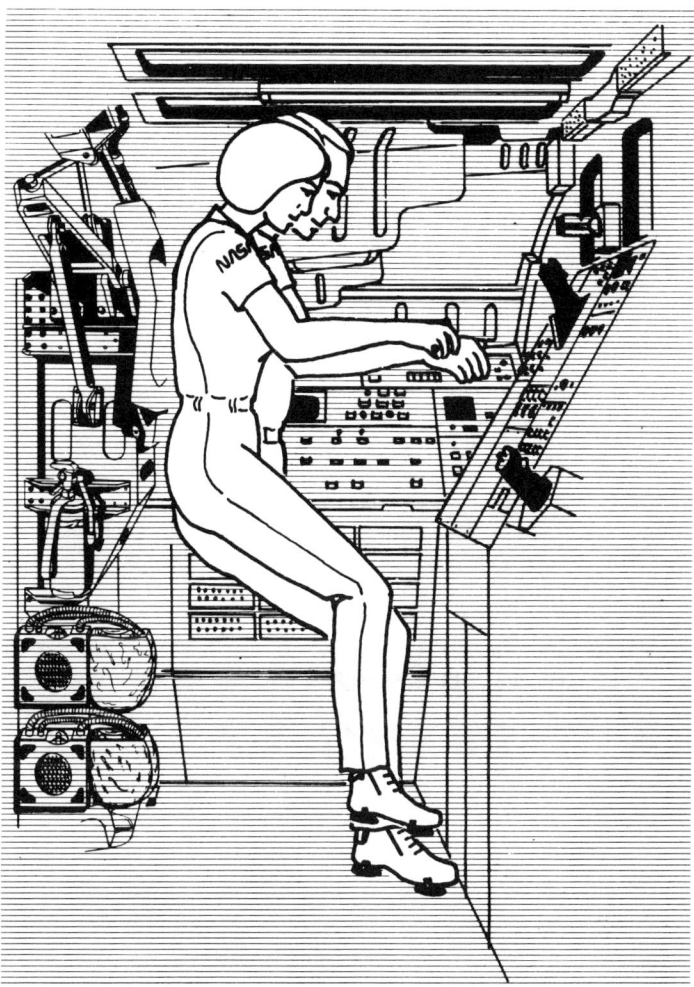

Figure 7. Crew size and neutral body position criteria for RMS Control Station.

Remote Manipulator System Control Station Design

The design that best illustrates the man-machine design using the findings of Skylab experiences is the Remote Manipulator System (RMS) control station. The RMS control station is located in the aft portion of the flight deck, on the port side. Figure 6 illustrates the location of the RMS control station.

Crew body size and neutral body posture were critical design factors in locating the overhead and aft windows, the display and controls, the closed circuit television, and the hand controllers. Figure 7 is a comparison of a 5th-percentile female and a 95th-percentile male positioned at the RMS control station. Gross assessment indicates that if eye point and limb mobility are held constant, a height adjustment is required. Other factors include reach and head movement for visibility. The RMS control station is an example of the erect workstation that was designed with criteria developed during previous space programs.

The major considerations in designing for weightless environments center on the factors that are different from earth-like environments. Notably, the change in body posture and the requirement to restrain a crewman in place emerge as significant design drivers which change the concept and the configuration of a workstation. Although subtle in terms of outward appearance, the changes implemented in these newly derived design standards will greatly facilitate the crewman's task in the weightless environment.

Paul C. Rambaut

6. Nutritional Criteria for Closed-Loop Space Food Systems

Introduction

As space missions become longer and longer, a point will be reached at which a system that will produce food at least partly from metabolic wastes will be necessary. A number of predictions have been made as to how long such a mission would have to be. In 1966, a report by the General Dynamics Company concluded that, for a two-year (or Mars-type) mission, regeneration of carbohydrates by physical-chemical means would allow the weight and volume of the food supply to be smaller than with stored food alone. In a parallel study, the Lockheed Missiles and Space Company came to a very similar conclusion. More recently Spurlock and his colleagues presented arguments to demonstrate that a crossover point occurs at mission lengths of about eight years, on the average _(1, 2, 3)_.

In general, the mass of stored food increases as a linear function of crew size and mission length. The point at which it becomes more economical to produce some food on board than to carry all of it from earth is dependent on many variables. Perhaps the most important of these are the mass and efficiency of the on-board food production system. Many schemes have been proposed for producing food from metabolic waste. The relative mass and efficiency of each of these systems account for the large variation in the estimate of the crossover point.

The physiological requirement is, of course, not for food but for nutrients (Figure 1). All potential on-board food production systems have in common the requirement to meet the nutritional needs of human beings living and working in space for extended periods. These nutritional needs are in fact the gross chemical specifications of any diet, whether put together from stored or recycled

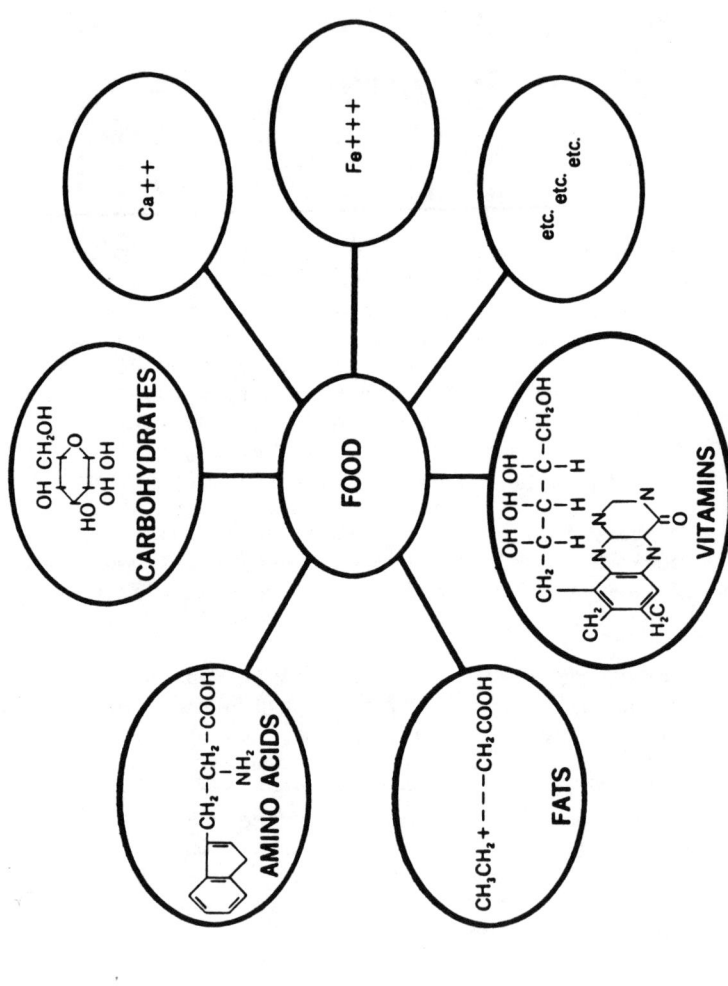

Figure 1. Food supplies nutritional needs through its nutrient constituents.

materials or from a combination of both. There is some
latitude for variation in these chemical specifications.
Consequently, it is important to recognize the upper and
lower limits set by that variation before much effort is
put into developing an on-board food production system.

The three Skylab visits, lasting 28, 59, and 84 days
in earth orbit, offered the first opportunity to draw any
firm conclusions regarding the nutritional requirements
of man in space. The purpose of this paper is to summarize
those requirements and point to their implications in
terms of the design of on-board food production systems.
In developing diets for the Skylab crews it was assumed
that nutrient requirements in space did not differ from
those on the ground. Our experience with Skylab showed
that this assumption was not completely correct.

Nutritional Requirements in Space

A determination of man's energy requirements in
space is crucial to the design of life support systems.
The need for energy largely determines the weight and
volume of the food supply and also establishes the
requirement for oxygen.

Examination of Skylab data revealed that the overall
energy consumed inflight was not statistically different
from that consumed on the ground. More careful scrutiny,
however, revealed that changes occurred as the flight went
on. The rate of energy utilization during the first month
inflight was significantly lower than the value obtained
during the month before flight; during the second month,
though, the value was significantly higher. This trend
continued into the third month as well. Linear-regression
analysis of the apparent increases in energy input rates
during each month of flight revealed that the trend toward
increasing rates of energy utilization was statistically
significant. The energy utilization rate (Figure 2) of
the Skylab crewmembers at the beginning of their 3-month
flight was 43.7 kcal/kg per day. This rate increased
approximately 1.6% per month. The average increase in the
"normalized" rate was about 3.7% per month (4).

The gradual elevation in the energy input inflight
indicates either an increasing energy output or a
decreasing metabolic efficiency. Quantitative measurements
of energy output were not made on a continuous basis.
However, while subjects were performing a standardized
level of work on a bicycle ergometer, the ratio of CO_2
produced to O_2 consumed did increase by an average of about

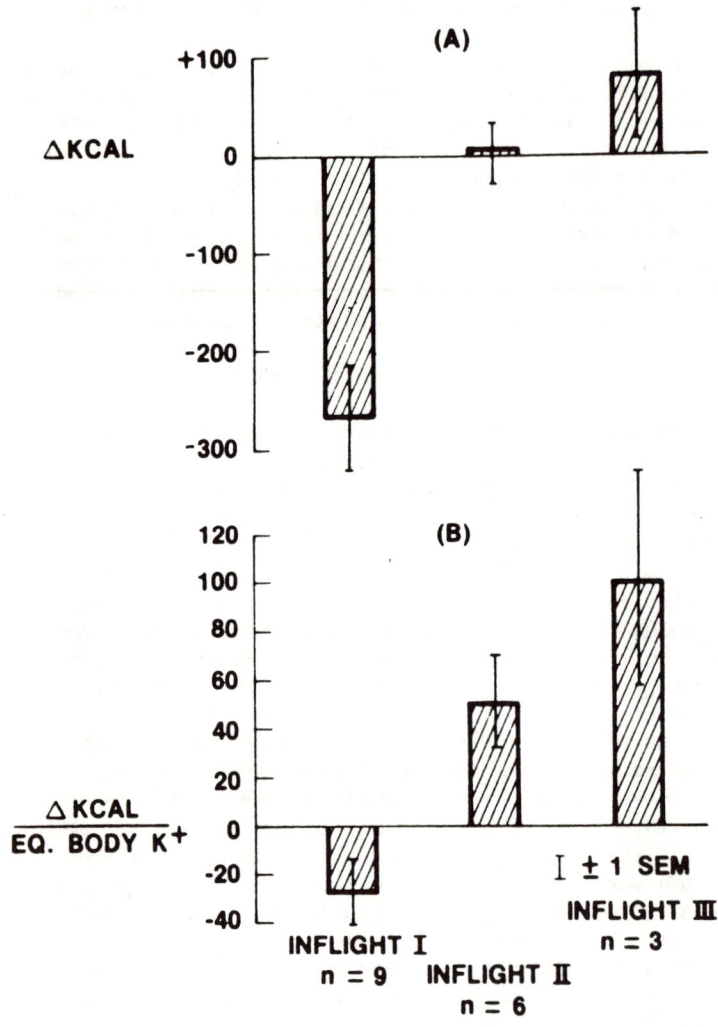

Figure 2. Calorie expenditure rates during the Skylab missions (by month). (A) Shift from preflight control in net energy utilization for each inflight period. (B) Shift from preflight control in normalized net energy utilization for each inflight period.

3% over the course of the flight. This observation, along with the apparent increases in thyroxine seen following recovery from the Skylab missions, lends some support to the concept that metabolic efficiency inflight may diminish. The loss may perhaps be attributable to the fact that proportionately more work was being demanded of a diminishing muscle mass -- in particular, of a diminished number of fast-twitch muscle fibers (5).

Fat

Although man requires some 40 specific organic compounds and minerals in his daily diet, their combined weight is only about 50 to 100 g. Relatively nonspecific sources of energy account for the remaining 400 to 500 g of daily food intake. In conventional diets, calories are derived from a mixture of energy sources with about 15% coming from protein, 30-50% from fat, and 40-60% from carbohydrate. As long as conventional foods are carried into space, it is unlikely that one or another of these energy forms would be excluded. However, foods can be selected to provide more or less energy from the three sources, and specially prepared food mixtures can be designed to incorporate any desired balance among these sources. If, on the other hand, food is to be regenerated wholly or in part, the regenerable products could result in extreme departures from the normal mixture. This would be the case, for instance, if the foods produced on board were relatively high in fat.

The minimum and maximum limits for fat must therefore be established. It turns out that only certain polyunsaturated fatty acids must be provided as such in the diet and, as far as is known, the requirements for these acids constitute the only dietary requirement for fat. Linoleic acid can fulfill all of this requirement. The minimum need can probably be met by a diet in which 1% of total calories is provided by linoleic acid. This amounts to about 3-4 g per day (6). As for the maximum limit, persons usually complain when fat intake exceeds about 150 g per day, although tolerance varies widely and those who are able to tolerate their initial discomfort often adjust to the diet.

When considering the type of fat to be included in the diet, attention must be given to the evidence associating dietary fat with cardiovascular disease. Present evidence suggests that the average chain length of fatty acids should be about 16 carbon atoms, with 1.2 g unsaturated for each g of saturated.

Carbohydrates

Although carbohydrates are the major energy source in conventional diets, carbohydrate need not be present as such. It can be formed from the glycerol moiety of fats or by gluconeogenesis from protein. The maximum tolerance for carbohydrate is unknown but varies both among individuals and with the specific carbohydrate given. The blood levels of lipids become undesirably high only when carbohydrate is increased beyond 85-90% of total calories.

Dietary carbohydrates of differing chemical structure are known to differentially influence work output in human subjects and to affect the central nervous system. These results may be related to reports that ingested carbohydrate elevates serum tryptophan and, hence, brain serotonin concentrations (7).

The carbohydrates are catabolized to the common intermediate trioses, D-glyceraldehyde and dihydroxyacetone, and then to carbon dioxide and water. It is probable that ingestion of substantial amounts of these intermediates would be well tolerated since they act biochemically in much the same manner as carbohydrate. This fact was the basis of earlier efforts to reduce carbon dioxide to trioses as the basis for space food production systems (8).

Amino Acids

The weight of essential amino acids needed to meet the requirement of healthy young men is about 5.5 g per day. The need for nonspecific amino acids is substantially larger, making up the remainder of the 35-45 g total protein required in the daily diet. It appears that this additional requirement can be met by a mixture of essential or nonessential amino acids or by an amino nitrogen source such as diammonium citrate. Generally, however, metabolic performance is found to be superior when protein, rather than free acids, is fed and when a mixture of nonessential acids is given, rather than ammonium compounds (9).

Digestibility varies among proteins. Usually, plant proteins are less digestible than are animal proteins. The digestibility of plant proteins can be improved by refining, whereby the protein is separated from the coarse, fibrous material. Mild heating or normal cooking often improves digestibility but intense heat is damaging. Because protein synthesis is an "all-or-none" reaction and the needed amino acid residues are delayed because of poor digestibility or other factors, the level of a particular protein required in the daily diet would be larger than that predicted from its amino acid composition alone. If digestibility is severely damaged and some protein is lost in the feces, the requirement is increased still further.

In Skylab and in missions prior to Skylab there was evidence for the occurrence of a net loss of nitrogen inflight (Figure 3). It is certain that part of this loss was due to hypocaloric nutriture and part to disuse atrophy *(10)*.

If insufficient fat and carbohydrate are included in the diet, protein will be catabolized for gluconeogenesis because the needs for energy and for carbohydrate intermediates usually have priority over the maintenance of tissue proteins. In future missions, once we have insured that energy needs are met, it will probably be wise to provide at least the recommended dietary allowance of high-quality protein. Whether or not more protein would be beneficial is a debatable point. Increased protein intake is not only unlikely to offset the protein loss due to muscle atrophy but also results in the loss of more water and reduced organic products in the urine.

Increases in dietary protein may expand blood volume and may ameliorate the physiologic effects of the body-fluid shifts that occur as a direct effect of the change from 1 to 0 *g*. A drop in plasma volume might be better tolerated if liberal amounts of protein are ingested preflight to deliberately expand plasma volume. Maintenance in-flight of the crew's habitual preflight protein intake would prevent the additional loss of fluid volume occurring with readjustment of labile tissue proteins.

The data available on amino acid requirements on the ground are very poor, particularly for females. Reliable information for amino acid requirements in spaceflight is non-existent. There is no evidence for alterations in amino acid metabolism inflight. The basis for the use of methionine, asparagine, and glutamic acid supplements by USSR crews in this regard rests primarily on pharmacological considerations.

In most societies relatively high protein diets are consumed whenever they are available. The relative importance of the psychological and the physiological causes for this clear preference is not known. In any event, the level and kind of protein has important implications in terms of satiety, acceptability, and the

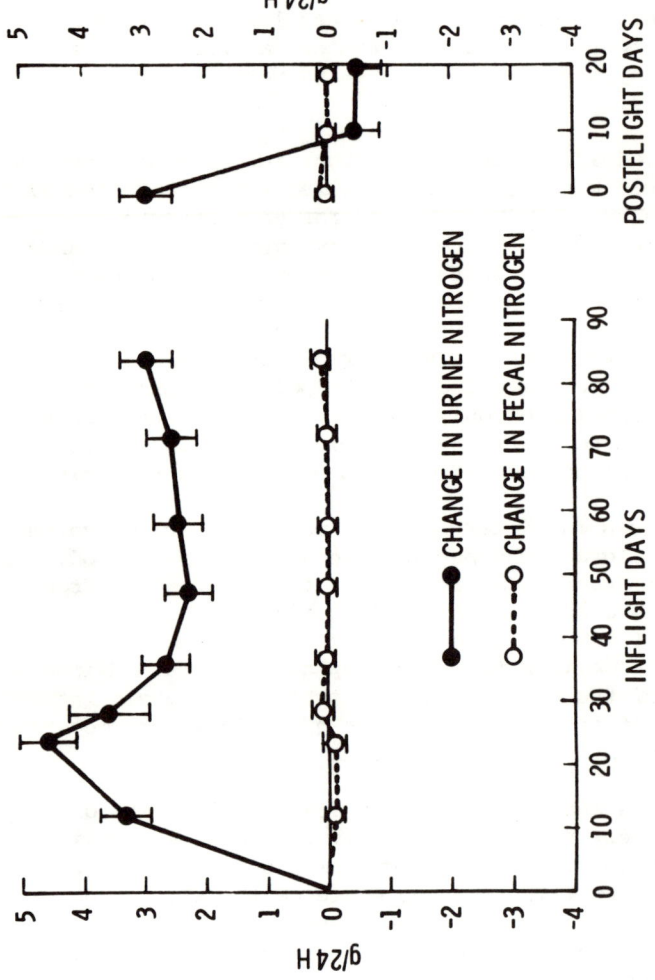

Figure 3. Change in urinary and fecal nitrogen as a function of Skylab flight duration.

general ability to keep people happy. Much effort is
needed to assess the psychological significance of problems
posed by diets low in protein.

The current recommended levels of dietary protein
make no allowances for a possible effect of physical
exercise. Vigorous exercise programs have been proposed
for astronauts to counter deconditioning. A resolution
of this effect of exercise on protein requirements is
important.

Macronutrient Elements

Figure 4 illustrates the composite calcium imbalance
sustained by all Skylab crewmen. Examination of the graph
reveals that calcium was lost exponentially as a function
of time inflight. Whereas an average of 50 mg was lost
on the 10th day, an average of 300 mg was lost on the 84th
day. An average of 25 g of calcium was lost from the
body during the 84-day flight. Assuming a total body
calcium content of about 1 kg, the pool was diminished in
size by approximately 2.5% (11).

When fecal and urinary losses of calcium are plotted
independently (Figure 5) two kinetically different
processes become evident. Urinary calcium rose rapidly
following launch and within 30 days achieved a level
approximately 100% above baseline. It remained at this
level for the remainder of the flight. Fecal calcium,
on the other hand, did not begin to rise for two or three
weeks following launch. Once fecal calcium began to
increase, it did so at a constant rate for the remainder
of the flight.

Apparently calcium is being mobilized from the
bones and is effecting a rise in the serum calcium and
phosphate concentrations. This change is accompanied
at first by a slight elevation and thereafter by a
sustained depression in serum parathyroid hormone at
about the same time that the mobilization of bone mineral
is occurring, less and calcium is being absorbed from
the gastrointestinal tract. Although normally sufficient
quantities of vitamin D-2 were being ingested, it is
possible that a deficiency was developing in the conversion
of vitamin D-2 to the 1,25-dihydroxycholecalciferol required
to affect the absorption of the approximately 800 mg of
calcium present in the daily ration.

Alleviation of the stress of weight bearing seems to
have a significant influence on skeletal metabolism.

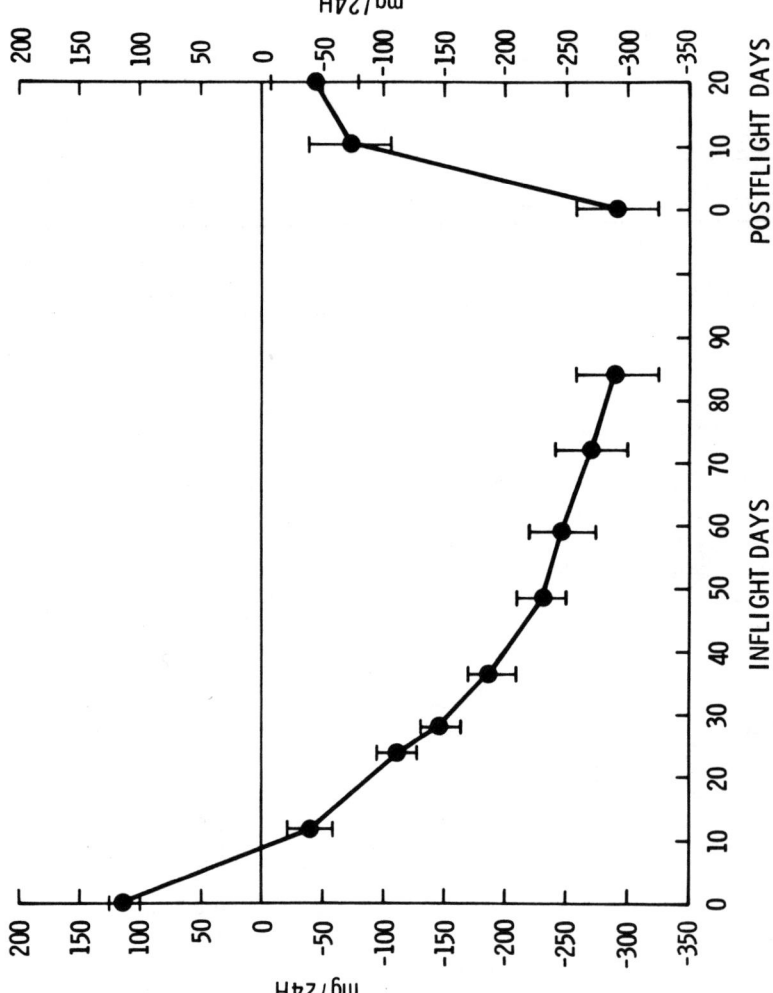

Figure 4. Daily calcium balance as a function of time during and after Skylab Earth Orbital Missions.

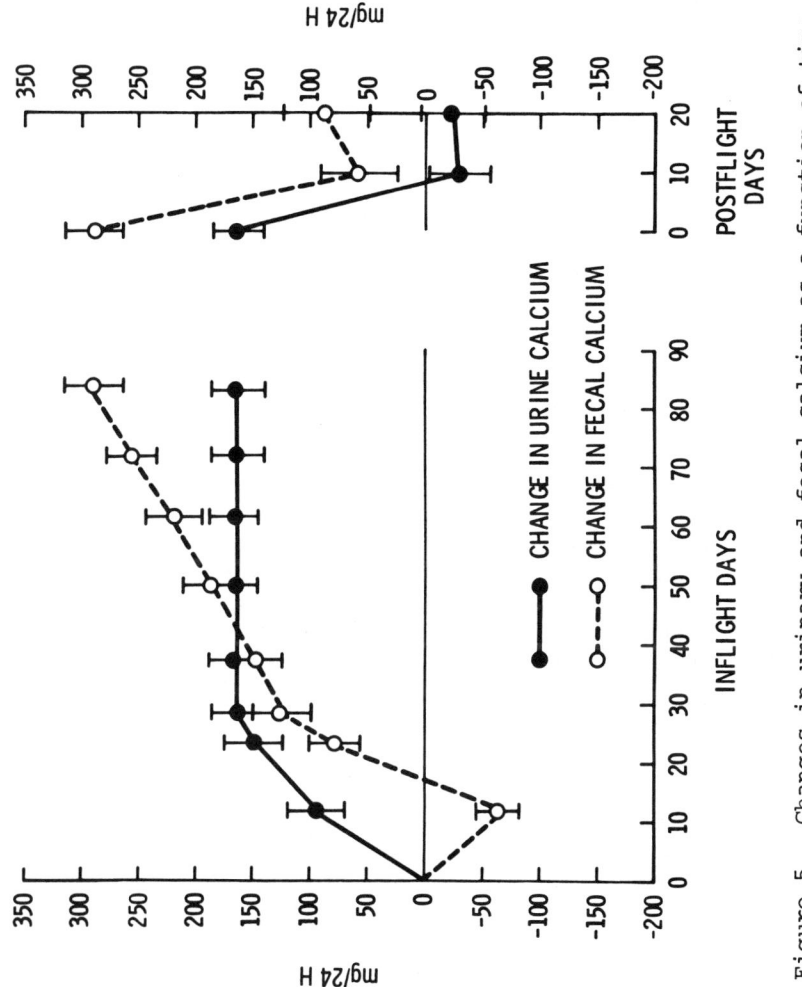

Figure 5. Changes in urinary and fecal calcium as a function of time during and after Skylab Earth Orbital Missions.

In rats placed aboard the flights of the Cosmos satellite series, investigators noted a severe depression of bone growth during flight (12).

The chain of events leading to this bone loss inflight remains unexplained. What biophysical or biochemical event is first affected by lack of gravity has not yet been identified. Hormonal, neural, circulatory, and mechanical factors are all undoubtedly involved at some point, and even piezoelectric factors have been implicated.

Several ways of ameliorating the bone losses have been studied. Protocols involving exercise, compressional suits, lower body negative pressure, and thyrocalcitonin administration have had little effect in moderating the losses induced by prolonged recumbency. However, changes in dietary intake, particularly reductions in protein and elevated intakes of diphosphonates, calcium, and phosphorus, are promising and point the way to the manipulation of nutritional intake as an important means of counteracting the adverse effects of prolonged weightlessness.

Micronutrient Elements

To state with certainty the precise quantities of the micronutrients that should be included in space diets is difficult to do, for various reasons. First, for many of the nutrients, the requirements of individuals living on Earth are not well-established. In fact, for some of the minerals, human requirements are known only qualitatively-or not at all. The fact that all these micronutrients are required in such small amounts and can be easily ingested in supplemental form makes less important a precise determination of their metabolic requirement. There are, however, some exceptions. One such exception is Vitamin D.

Vitamin D is essential for the maintenance of calcium homeostasis. It is essential for the proper absorption of calcium from the intestinal tract and for the proper action of the parathyroid hormone on bone. In the absence of ultraviolet radiation the ingestion of small amounts of Vitamin D, possibly 400 I.U. per day, is necessary for adults.

The increased excretion of calcium derived from bone probably cannot be ameliorated by simply administering

Vitamin D. However, adequate amounts of this vitamin must be present to maintain inflow of calcium through the gastrointestinal tract.

The range between minimal or optimal and toxic is extremely small in the case of Vitamin D. Some individuals appear to be so sensitive that Vitamin D dosages of as little as 1000 to 3000 I.U. per day may be toxic. The toxicity of Vitamin D leads to hypercalcemia. This calcemia is probably related to the increased mobilization of calcium from bone. In view of the possibility that weightlessness could lead to increased mobilization of calcium from bone, to increases in urinary calcium, and possibly to hypercalcemia, the potential toxicity of Vitamin D is an important consideration.

The trace minerals for which a requirement is currently recognized are cobalt, copper, chromium, iodine, iron, manganese, molybdenum, selenium, zinc, and possibly aluminum. In prolonged space flight the presence of these minerals must be assured. While there is no evidence at present to suggest that mineral requirements would be modified by space flight, there must be a constant awareness that the space environment may provide totally unexpected surprises as man attempts to remain in that environment for longer and longer periods of time.

It is commonly assumed that all human nutritional factors are known. But so far no actual test has been undertaken of an individual fed for long periods of time on highly purified materials. The possibility that other trace elements or organic compounds may be essential for life must be considered. The experiences of those faced with problems of hyperalimentation may be very useful in this connection.

In closed systems, there exists the added danger of toxicity at high levels of nutrient elements. There is often a narrow tolerance between the minimum requirement level and the toxic level. A better grasp of requirements and tolerances and an efficient monitoring system for the micronutrient elements will be required.

Food Production Considerations

Several suggestions have been made for the production of foods inflight (Figure 6). Among these are methods employing photosynthetic organisms, tissue cultures, nonphotosynthetic organisms, and cultures of higher animals. In contrast to the use of living tissues, reprocessing

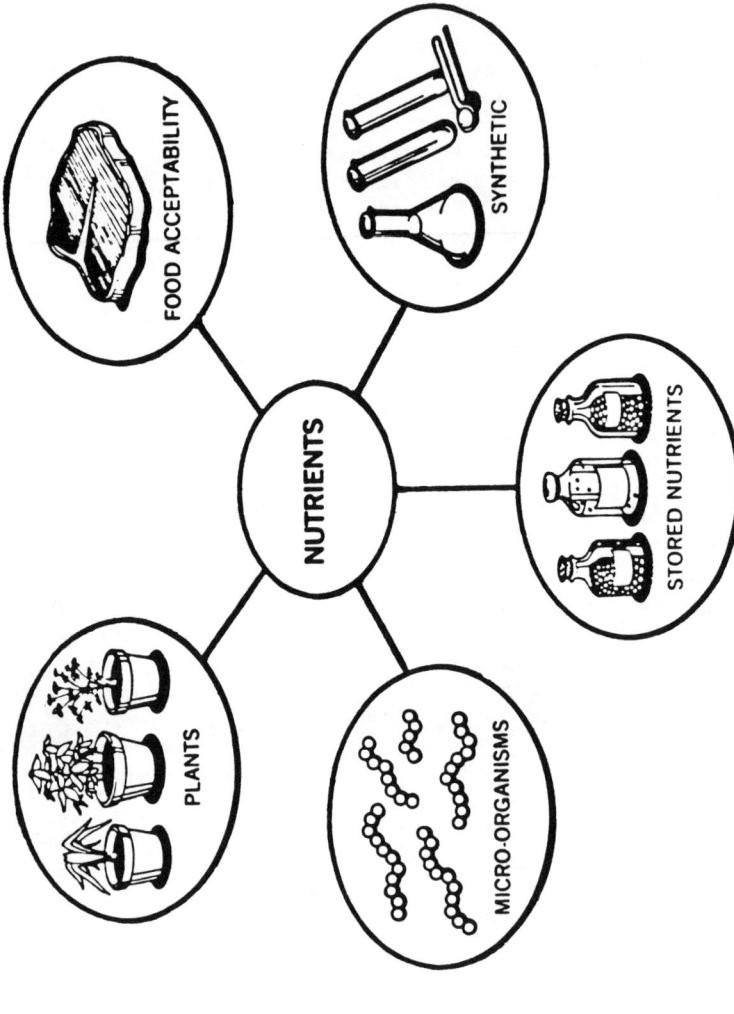

Figure 6. Alternative methods of producing food inflight.

of waste materials by chemical treatment and the actual synthesis of nutrients have also been suggested.

The photosynthetic process is attractive because it is the conventional way of closing the Life Support System. Through this process there are produced compounds at least as reduced as H_2 and at least as oxidized as O_2. The end result in the green plant is the evolution of O_2 and the reduction of CO_2 to organic materials. The energy to power this system is to be supplied in the form of light (Figure 7).

Plants vary tremendously in their content of the three major groups of organic compounds. A large part of this variation arises from two simple biological principles: (1) that working machinery is largely protein and (2) that larger organisms have proportionately less working machinery and proportionately more skeletal material. In plants, sketetal material is cellulose, which is not digestible by the human.

As we proceed from a microscopic alga to the small duckweed, to the large corn plant, to the giant sequoia, there is a progressive reduction in the proportion of working machinery and a progressive decrease in manageability of the plant. In the past, algal cultures have had the most extensive investigation as a potential source of food in space, but the technological problems of using this material as a food source have not yet been solved and currently much more emphasis, both in the USA and in the USSR, is being placed on the use of higher plants.

In fact, so much effort has been and is being made to evaluate photosynthetic systems that the theoretically greater advantages of chemical synthesis of specific nutrients has attracted relatively little attention.

If carbohydrates or fat could be synthesized on-board, it is possible that less than 10 percent of the total food requirements of the astronaut would have to be carried along. This stored material would supply only those components of the diet that are difficult to synthesize, such as essential amino acids, unsaturated fatty acids, and vitamins. Even these might be provided by incorporating into the diet relatively small amounts of material from a photosynthetic system.

The principal advantages of physicochemical methods of food production are: high energy value of the substances formed; their high nutrient efficiency, low weight and

Figure 7. Concept for a Spacelab experiment with an intensive agriculture system.

volume, good assimilability, and ease of preparing food from them; the possibility of automatic regulation and control; and the relative independence of physicochemical processes from such significant spaceflight factors as weightlessness and ionizing radiation. A major disadvantage is that such synthetic foodstuffs are unappealing.

These products, however, can be given a more appetizing taste and odor by special food additives or by additional processing. Powdered substances can be converted into gelatins, which are more convenient to eat, with the use of synthetic polymer compounds.

We must re-examine the prevailing assumption that humans require a diet of great variety. In many parts of the world, people live on diets consisting of only a few types of food with no apparent ill effects, provided their basic nutritional requirements are satisfied. Experimental evidence from many sources shows that individuals can now be kept on a simple nutrient source for many months without suffering ill effects. There is also a good possibility that a formula diet of greater acceptability than those used today will be developed.

The sensory properties of food do not alone govern acceptability. Acceptability is an evasive concept. Intrinsically, food tastes neither good nor bad. The taste of food is largely determined by conditioning and by internal chemistry. When it was introduced, powdered milk had low acceptability among American soldiers on bases. To them it tasted bad. The same powdered milk, however, had a very high acceptability among similar soldiers who were starving in a prison camp. To these men, powdered milk had a delicious taste. The difference was not in the powdered milk, but in the biochemical status of the starving men.

Skylab crews undertook a controlled study to determine whether there were actual changes in the response to taste or aroma during weightless flight. The results of the tests gave no evidence that there was any change in the ability of any crewmember to identify aromas. There was also no evidence in any of these data to suggest that weightlessness is a significant factor influencing food flavor. Yet food "taste" and acceptability have been reported to change inflight by both American and Soviet crews.

Probably the ideal space food lies somewhere in between a simple formula diet and a diet of great variety.

In any event we should be wary of technologically expedient solutions. We could, if we are not careful, provide food upon which space crews could survive a lot longer than they would care to live. For whatever purpose or by whatever means we decide to develop food production systems, it is clear we are embarking upon a very significant step in human evolution. Our ability to produce food, first from our own waste material and perhaps one day from the substance of other worlds, is the most fundamental measure of our progress in the conquest of space.

References and Notes

1. G.L. Drake, C.D. King, W.A. Johnson, E.A. Zuraw, "Study of Life Support Systems for Space Missions Exceeding One Year in Duration," in *The Closed Life Support System*, H.P. Klein, Ed. (SP-134, NASA/Ames Research Center, Washington, D.C., 1967), p.1.

2. R.B. Jagow and R.S. Thomas, *ibid.*, p.75.

3. J.M. Spurlock and M. Modell, *Technology Requirements and Planning Criteria for Closed Life Support Systems for Manned Space Missions: Final Report* (Contract Number NAS-W2981, NASA/Ames Research Center, Moffitt Field, Calif., 1978).

4. P.C. Rambaut, C.S. Leach, J.I. Leonard, "Observations in Energy Balance in Man During Spaceflight," *Am. J. Physiol. 233* (1977), pp. R208-R212.

5. C.S. Leach and P.C. Rambaut, "Biochemical Observations of Long Duration Manned Orbital Spaceflight," *J. Am. Med. Wom. Assn. 30* (1975), pp. 153-172.

6. D.H. Calloway, "Dietary Components That Yield Energy," *Environ. Bio. Med. 1* (1972), pp. 175-186.

7. W.J. Shoemaker and F.E. Bloom, "Effect of Undernutrition on Brain Morphology," in *Nutrition and the Brain*, R.J. Wurtman and J.J. Wurtman, Eds. (Raven Press, New York, 1977), p. 148.

8. H. Forster, "Comparative Metabolism of Xylitol, Sorbitol and Fructose," in *Sugars in Nutrition*, H.L. Sipple and K.W. McNutt, Eds. (Academic Press, New York, 1977), p. 259.

9. D.M. Hegsted, "Proteins in Space Nutrition," in *Nutrition in Space and Related Waste Problems*, M.G. Del Duca, Ed. (SP-70, NASA, Washington, D.C., 1964), p. 135.

10. P.C. Rambaut, M.C. Smith, C.S. Leach, G.D. Whedon, J. Reid, "Nutrition and Responses to Zero Gravity," *Federation Proceedings 36*, (1977), pp. 1678-1682.

11. P.C. Rambaut, C.S. Leach, G.D. Whedon, "Prolonged Weightlessness and Calcium Loss in Man," submitted to *Am. J. Clin. Nutr.*

12. E.R. Morey and D.J. Baylink, "Inhibition of Bone Formation During Space Flight," *Science 201*, (1978), pp. 1138-1139.

_____ *Michael Modell, Jack M. Spurlock*

7. Rationale for Evaluating a Closed Food Chain for Space Habitats

Introduction

Forecasts for future manned space missions, beyond those typified by the Space Shuttle and Spacelab flights, as now defined, include various mission concepts that require teams of many human participants working and living in space for extended periods of time. Most of the concepts under discussion for the near term (1980-2000) involve space industrialization objectives of some type, and preliminary design and feasibility studies are increasing in number and scope. These potential missions may involve the order of tens of people in space for a period of months. In the early twenty-first century, it is conceivable that missions might involve hundreds of people in space for a period of years. With closed-environment life support technology yet in its infancy, it is timely to begin studies of technology requirements and design concepts to provide advanced life support systems that would be appropriate for such missions. The purpose of this paper is to present an overview of some major factors that will be associated with the design of these systems for operation in the space environment.

Life support systems for manned space missions provide the crewmembers with food, potable water, a habitable atmosphere, and waste management, as well as personal-hygiene facilities. The scope of manned missions that have

---

The analyses which provided the basis for this paper were sponsored in part by the Office of Life Sciences, Headquarters, National Aeronautics and Space Administration, on a contract with the Society of Automotive Engineers. This sponsorship is gratefully acknowledged. In addition, the authors appreciate the data that were supplied by members of the Bioenvironmental System Study Group as part of the work on that contract.

been flown to date has permitted the use of non-regenerative methods of life support. That is, food, water, and the breathable atmosphere were supplied from material that was transported into space from the earth and consumed as needed, with no attempt to regenerate those materials from the waste products that resulted from their consumption. The wastes were either discharged from the spacecraft or treated and stored on board. This approach represents the "open" mode of meeting life support requirements. The inputs required to support a person in space are shown in Table 1.

Table 1. Inputs required to support a person in space (3).

Inputs (kg. per person)	Day	Year	Lifetime
Food (dry)	0.6	219	15,300
Oxygen	0.9	329	23,000
Drinking water	1.8	657	46,000
Sanitary water	2.3	840	58,800
Subtotal	5.6	2,045	143,100
Domestic water	16.8		
TOTAL	22.4		

Over the course of a year, the average person consumes 3 times his body weight in food, 4 times his weight in oxygen, and 8 times his weight in drinking water. Over the course of a lifetime, these materials amount to over 1000 times an adult's weight.

For long-term space missions, it will undoubtedly be necessary to replenish or recycle supplies on board. The quantities of supplies required increase with the duration of the flight, and the cost of including those supplies is proportional to their weight at launch. At some break-even point it will begin to be less costly to send up the equipment needed to recycle the supplies from wastes than to use up the provisions on a once-through basis. For example, wash water is relatively easy to recycle because the contaminant level is only 350 ppm (1). A number of processes have been investigated for wash-water recycle, including systems based on multifiltration, ultrafiltration, reverse osmosis, and vapor-compression distillation. Based on currently available technology, the break-even point is about a week for a crew of six people (2). Therefore, wash-water recycle systems will undoubtedly be flown at an early stage.

The next level of recycle, beyond wash water, would involve closure of the entire water loop. Assuming that 90% of the wash water is recoverable, a crew of six in space for six months would require over 6,000 kg of water from stores for drinking, sanitary flush, and wash water make-up. This water could be recovered by processing the combined toilet waste and wash water concentrate *(3)*. Systems for such waste treatment have undergone varying degrees of research and development in previous NASA-sponsored efforts. Both wet and dry oxidation systems have been built and tested on a demonstration scale. In wet oxidation, an aqueous slurry is treated at moderate temperatures (150-250°C) under high pressures (1200-2200 psig) *(4)*. In dry oxidation, aqueous slurries are first concentrated by evaporation of water, and then the concentrate is incinerated *(5)*. Water vapor and off-gases from the incinerator are passed through catalytic oxidizer beds to completely convert organic contaminants to CO_2 and H_2O. For systems of this type, an equivalent system weight at launch of 6,000 kg will be sufficient to recycle the entire waste water for a crew of six *(3)*. Therefore, the break-even point for closing the water loop is on the order of six months.

A number of systems have also been investigated for recycling oxygen from carbon dioxide *(6,7)*. These methods involve separation of CO_2 from air, followed by reaction of CO_2 to produce O_2 and carbon by-products (methane, in the Sabatier process, or solid carbon in the Bosch process).

The combination of these physico-chemical techniques can provide for closure of all essential life support requirements, with the exception of food. The schematic of such a partially closed environment life support system (PCELSS) is shown in Figure 1.

Completely Closed Systems

For missions longer than a year in duration, with tens to hundreds of people, it is conceivable that it might be more efficient to close the material loop completely. To do so would require processing wastes into a form that could be incorporated back into the food chain. Before attempting to estimate the break-even point for a completely closed system, however, we must first determine whether such an enterprise is feasible. To what extent is current technology available to provide for a completely closed system? Several NASA-sponsored summer studies have addressed this issue, with the conclusion that - at least on paper - it is within the realm of current technology to close the loops with regard to the material balances of the major elements *(8,9)*. That is not

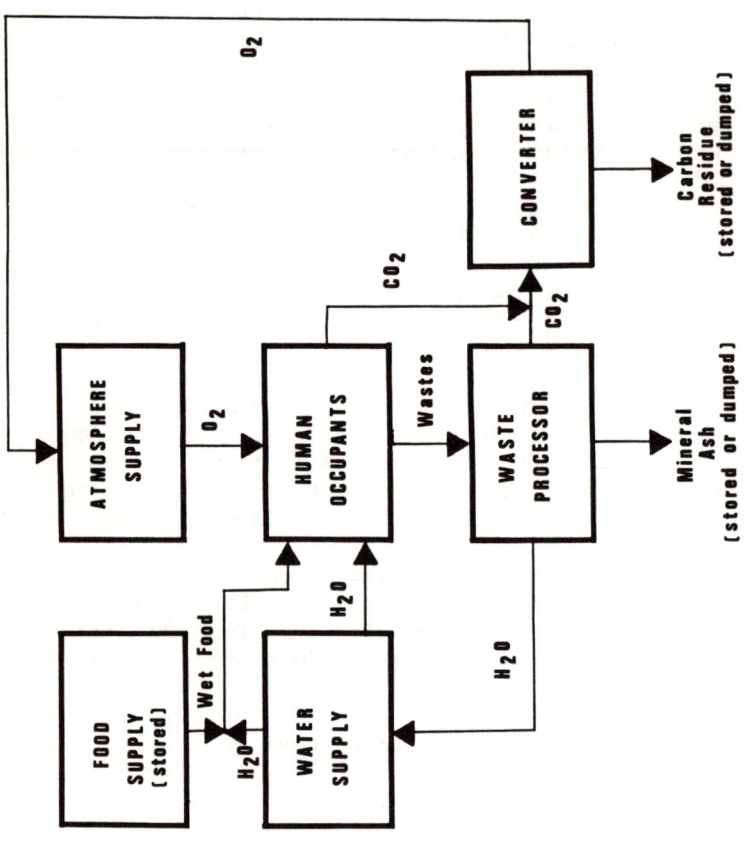

Figure 1. Example of a partially closed life support system.

Evaluating a Closed Food Chain 137

to say that such a closed system will be stable for a long duration. There are a large number of unknowns that must be addressed before the viability of a completely closed environment can be determined.

Considering the relative states of food-production technology, the existing data base is far more extensive for conventional (i.e., agricultural) methods than for the relatively undeveloped field of chemical synthesis of protein and carbohydrates for human consumption. Therefore, conventional methods currently provide the least speculative basis for the analysis of design requirements for food-chain closure. On this basis a completely closed system will involve at least three major components: humans, plants, and animals (the inclusion of aquatic life can also be envisioned). The inputs and outputs of these three subsystems are shown in Figure 2. Closing the system requires the conversion of each of the outputs back into inputs. Since all of the carbon in human food originates either directly or indirectly (i.e., through animal consumption) from vegetation, all of the carbon in the outputs from the three subsystems should be converted back to carbon dioxide so that it can be photosynthesized back into vegetation. In this manner, we can simultaneously satisfy the recycle requirements for oxygen and hydrogen. For example, if the average carbon, hydrogen, and oxygen composition of vegetation is represented by the formula CH_mO_n, then the stoichiometry of net plant growth (photosynthesis less respiration) is:

$$CO_2 + \frac{m}{2} H_2O = CH_mO_n + (1 + \frac{m}{4} - \frac{n}{2}) O_2$$

A portion of the edible plant output is oxidized back to CO_2 and H_2O directly by human and animal metabolism. The remainder of the edible portion is partially oxidized to products which appear in urine, feces, perspiration, and exhaled breath. If these partially oxidized products, together with the nonedible portion of the plants, are subsequently oxidized to CO_2 and H_2O, then the net effect of human and animal metabolism plus waste oxidation is the reverse of the net photosynthesis stoichiometry. Thus, incorporation of the oxidation process for complete conversion of organic wastes to CO_2 and H_2O simultaneously closes the carbon, oxygen, and hydrogen recycle loop.

In addition to CO_2 and H_2O, plants require on the order of a dozen essential nutrients for proper growth (N, K, P, Ca, Mg, S, Fe, Cl, Zn, Mn, B, Mo, and Cu) *(10)*. Since the entire harvest eventually becomes waste in one form or another, all of the plant nutrients are present in the

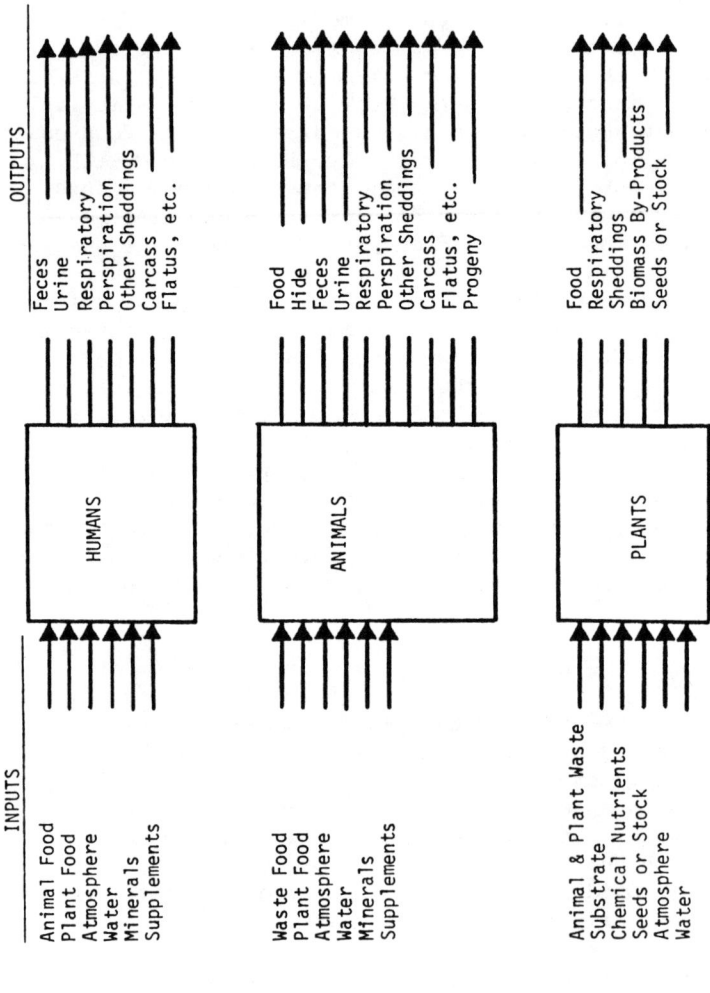

Figure 2. Inputs and outputs for a generalized closed-ecology life support system.

effluent of the waste oxidizer. Thus, separation of plant
nutrients from the waste effluent is the last essential step
in closing the loop.

A schematic for an entirely closed system is shown in
Figure 3. Liquid wastes from human and animal metabolism
could be concentrated by evaporation or vapor-compression
distillation in the unit labelled Primary Liquid Waste
Treatment. The concentrate, along with solid waste from
humans and animals and slurry from the wash-water recycle
system, could be fed to the Solid Waste Oxidizer, which
might be an incinerator. Air for oxidation could be withdrawn from the human and animal compartments. This process
would simultaneously sterilize the air and remove organic
contaminants from it. The off-gas from the Solid Waste
Oxidizer would contain acid gases (e.g., sulfur and nitrogen
oxides), which could be removed in an Acid Gas Scrubber
before passing the CO_2-rich atmosphere to the phytotron.
The ash from the Solid Waste Oxidizer would contain sodium
chloride and most of the elements required for the plant
nutrient medium. The metals in the ash would be present as
oxides, most of which have limited solubility in water.
Sodium chloride might therefore be extracted selectively by
mild aqueous treatment. The remaining oxides could be
solubilized by extraction with the condensate from the Acid
Gas Scrubber.

Rationale for Evaluating Closed Systems

The system shown in Figure 3 is one of many that might
be proposed for a closed life-support system. At the
present time, not enough is known to support the identification of optimum choices. Therefore, it will first be
necessary to identify the various technical approaches that
are possible and potentially promising, and the considerations involved in including these approaches in many
possible combinations to achieve effective and stable
functioning. The scenario-analysis approach is recommended
for this initial step of identifying and comparing closure--
and, therefore, research--options. It involves the
formulation and assessment of a variety of representative
schemes, or scenarios, for system closure (partial as well
as complete, to satisfy a variety of potential mission
applications). Each scenario differs from the others in the
choice of diet, together with the choice and internal interfacing of processing components. Several auspicious choices
of diet and processing techniques which can provide the
basis for scenario formulation are listed in Table 2.

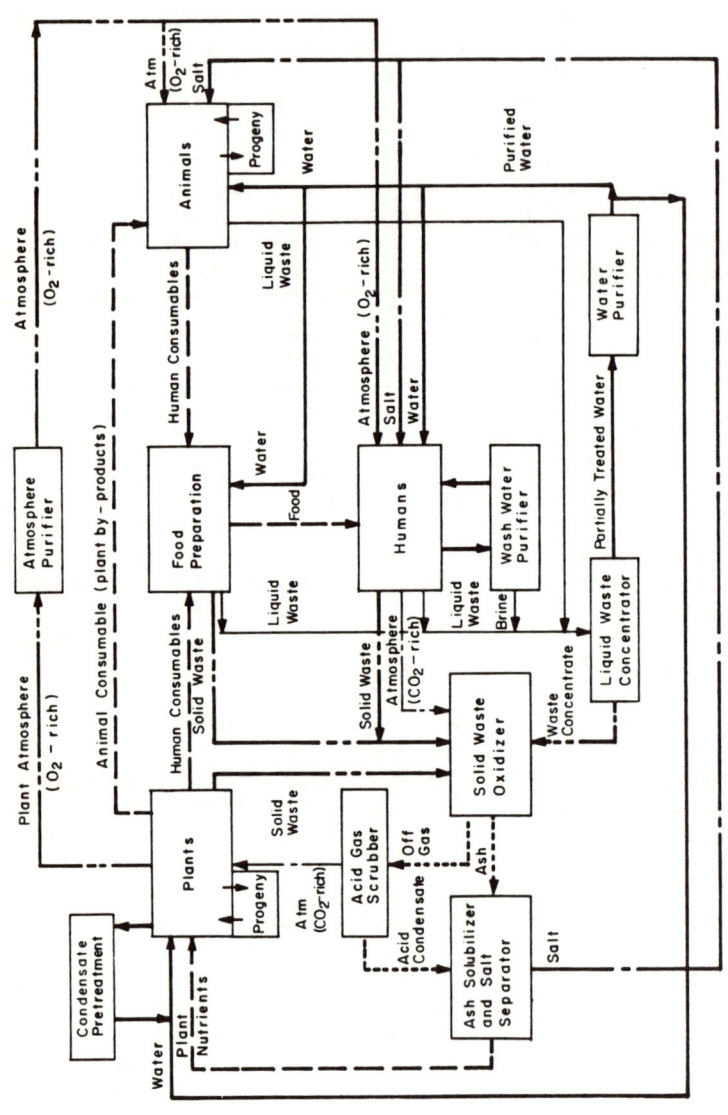

Figure 3. Schematic of a completely closed system.

Table 2. Typically chosen options for scenario components.

Diet	Food Source	Waste Processing*
Conventional Terrestrial	Produced in Space	Physicochemical
Abundant variation	Agriculture	Microbiological
Austere variation	Aquaculture	Agriculture
Vegetarian (only)	Chemical Synthesis	Aquaculture
Unconventional	Produced on Earth and Transported to Habitat	

* Includes atmosphere regeneration and water reclamation.

A very effective procedure for scenario formulation is shown schematically in Figure 4. The first step in this recommended procedure is the selection, from among the various possibilities, of the diet scenario, which specifies the types and quantities of food that will satisfy dietary requirements. Next, the agricultural or non-agricultural processes are to be included in the scheme, plant-growing requirements are determined first, and the amount of plant material that will not be consumed by humans is estimated from the agricultural characteristics of the plants. At this point, requirements for feed and resources to produce any animal-derived food components in the diet scenario are determined, with emphasis on maximum possible direct utilization of plant material that will not be consumed by humans.

In the next step, processing schemes must be selected for the conversion of wastes (from human and agricultural or non-agricultural food-producing components) into regenerated atmospheric constituents, water, and other appropriate inputs for humans and the food-producing components of the system (e.g., nutrients). The scheme is sufficiently defined at this point that the general types and sizes of processing equipment, facilities, and resources (e.g., water and energy) needed to accomplish the previous steps can be estimated. Typical examples of the latter include food processing, preparation, or synthesis equipment, storage facilities, refrigeration, enclosures for animals, crop cultivation and harvesting equipment, microbiological processing equipment, and waste or contaminant oxidizers. Finally, the stability characteristics of the candidate closed system, as well as requirements for instrumentation and control apparatus to stabilize the system, must be

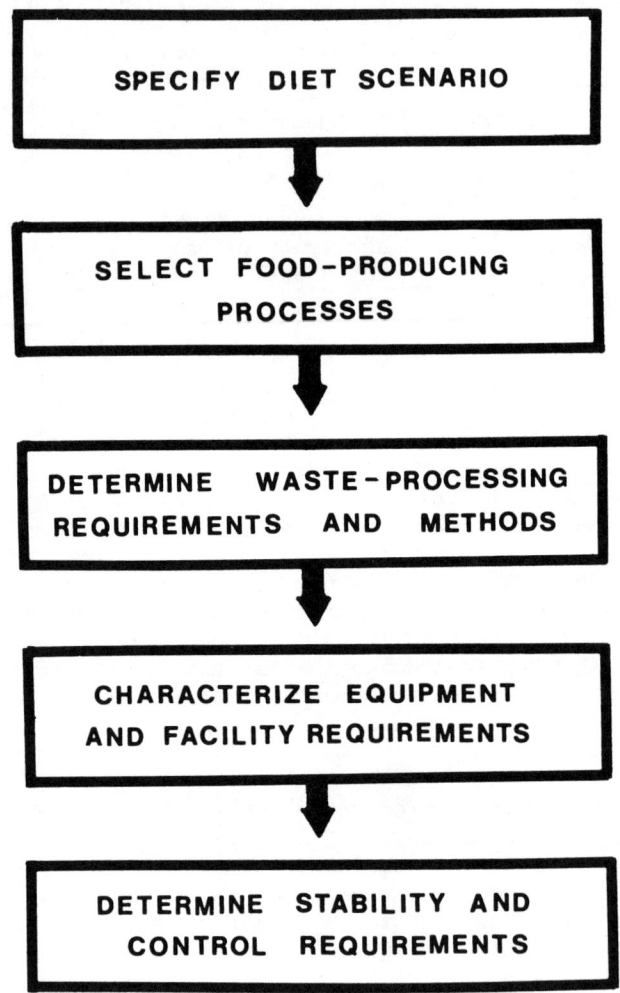

Figure 4. Procedure for scenario formulation.

assessed. This provides a basis for analyzing the system's sensitivity to perturbations at its internal and external interfaces.

The recommended methodology requires the consideration of a sufficient number of alternative scenarios to thoroughly cover the range of possibilities for diets, food sources, waste processing, air and water regeneration, and combinations thereof, in non-terrestrial as well as terrestrial operating environments, and covering partial as well as complete closure. For some of these scenarios, it will be very evident that the scheme will be impractical--or, at least, highly predisposed to functional and stability problems. For others, the tradeoffs will not be clear, based on currently available data.

After several reasonable scenarios have been formulated and analyzed for sensitivity to variations or perturbations at their interfaces, as well as for their dependence on the reliability of assumptions, they can be compared. Tradeoff analyses and cost-risk-benefit evaluations are suitable for this comparison procedure, and can provide an effective basis for selecting the most promising scenario options for further consideration. Potential research and development costs and time requirements should be estimated for each scenario. Where these estimates are very sensitive to the accuracy of technical estimates and data bases that are not very reliable at present, research requirements to refine these estimates should be defined. Methodology for tradeoff analyses can be adapted from techniques that have already been developed specifically for life-support systems components (3).

For the scenario options that become identified as most promising by the comparison procedure, high-priority pacing research needs can be determined from a thorough study of potential problem areas for each scenario, including anticipated performance limitations in optimum as well as non-optimum modes of operation. This information can be derived from the sensitivity analyses that were performed earlier for each scenario. Highest research priority should be assigned to investigations that show the most attractive cost-effectiveness potential and the most universally valuable potential results.

As the data base improves through the results of the initial high-priority research investigations, the tradeoff comparisons should be updated regularly. The results of these comparisons can provide a basis for (1) determining cost effectiveness of the research investment to date;

(2) redirecting future efforts; (3) predicting and comparing the time of readiness for use (i.e., technological maturity) for the various system options; and (4) identifying promising new system-scenario options based on combinations of particularly attractive components or subsystems to form hybrid groupings. This iterative systems engineering procedure can provide an important contribution to the effective management of the multidisciplinary technology-development programs that will be required in the development of successful regenerative life-support systems for space habitats.

References and Notes

1. D.F. Putnam and G.V. Columbo, *Experimental Study of the Constituents of Space Wash Water*, Contract No. NAS2-8239, NASA CR-137735 (Umpqua Research Co., Myrtle Creek, Ore., 1975).

2. D.F. Putnam, *Development Assessment of Wash Water Reclamation*, Contract No. NAS2-8239, NASA CR-137934 (Umpqua Research Co., Myrtle Creek, Ore., 1976).

3. J.M. Spurlock, M. Modell, D.F. Putnam, L.W. Ross, J.N. Pecoraro, *Evaluation and Comparison of Alternative Designs for Water and Solid-Waste Processing Systems for Spacecraft*, Final Report, NASA Contract No. NASw-2439 (Society of Automotive Engineers, Inc., Warrendale, Penn., July 1975).

4. R.B. Jagow, "Development of a Spacecraft Wet Oxidation Waste Processing System," ASME Paper 72-ENAv-3, Presented at ASME Environmental Control and Life Support Systems Conference, San Francisco, Calif., August 1972.

5. J.D. Schelkopf, F.J. Witt, R.W. Murray, *Integrated Waste Management-Water System Using Radioisotopes for Thermal Energy*, Document No. 74 SD 4201 for AEC Contract No. AT(11-1)-3036 (General Electric Co., Valley Forge, Penn., May 1974).

6. R.C. Reid, *University Role in Astronaut Life Support Systems: Atmospheres*, NASA CR-1552 (Massachusetts Institute of Technology, Boston, 1970).

7. P.D. Quattrone, "Spacecraft Oxygen Recovery Systems," *Astronautica Acta 18*, 4 (1973), pp. 261-271.

8. R.D. Johnson and C. Holbrow, Eds., *Space Settlements: A Design Study* (NASA SP-413, National Aeronautics and Space Administration, 1977).

9. J.M. Spurlock and M. Modell, "Systems Engineering Overview for Regenerative Life Support Systems Applicable to Space Habitats," NASA Summer Study, 1977 (in press).

10. R.J. Downs and H. Hellmers, *Environment and the Experimental Control of Plant Growth* (Academic Press, New York, 1975).

_____ Marcus Karel

8. Problems of Food Technology in Space Habitats

Recent developments in space technology have led to a renewed interest in the possibility of maintaining space colonies, or at least manned bases in space, in which groups of people can function during multi-year missions. Some of the design studies have in fact dared to consider the possibility of permanent settlements, or habitats, being initiated in the present century *(1)*. Consideration of life support systems becomes imperative in planning such ventures. One of the essentials of life support is, of course, food. The present paper is devoted to the discussion of research which will be necessary in order to plan the food technology required for space habitats.

In attempting to forecast an uncertain technological future it is useful to consider the fate of previous prophecies; and Jules Verne is of course <u>the</u> prophet insofar as space prophecies are concerned.

In his prediction of a voyage to the moon he included the menu shown in Table 1*(2)*. Table 2 shows the actual menu of astronaut Cernan on Apollo 17 *(3)*. As we can see, Verne was not far off the mark.

In thinking about longer space sojourns, Verne was less imaginative. In his functional *Voyage to the Moon*,

---

This work was supported in part by the Ames Research Center, NASA, whose support is gratefully acknowledged. In addition the author is grateful, for discussions leading to the evolution of the concepts shown in Figure 1, to Dr. J. Peter Clark of ITT Continental Baking Company, Dr. R. Sauer of the Johnson Space Center of NASA, and Dr. P. Singh of the University of California at Davis. He is also indebted to Dr. N. Heidelbaugh of Texas A&M University and D.I. Saguy of M.I.T. for their comments and suggestions on the subject of this paper.

he had the space travelers take along conventional agricultural tools expected to be useful in agriculture on the moon.

In yet another book (4), Verne envisioned a large group of people on a comet, but he providentially allowed the comet to have dragged along with it several hundred square miles from the Mediterranean area, a ship full of stores, an atmosphere and gravity in acceptable range, and even a cheap energy source in the form of an extinct volcano. The only reference to food processing problems in space was the lower boiling temperature of water and associated problems of cooking eggs properly--and in that discussion Verne makes a major thermodynamic error which should be detectable by any high school senior!

Table 1. Menu of a meal consumed on the way to the moon: fiction (2).

Liebig dehydrated beef soup base

Hydraulically compressed beef steak

Preserved vegetables

Bread and butter

Tea and French wine

Table 2. Menu of a typical meal consumed on the way to the moon: reality (3).

Apollo 17 (Cernan)
Day 5, Meal C

Rehydrated potato soup

Thermally stabilized beef and gravy

Rehydrated chicken stew

Rehydrated peach ambrosia

Intermediate moisture gingerbread bites

Rehydrated citrus beverage

Moving on to more modern prophets, in 1958 *Food Technology* published a symposium on problems of feeding humans in space (5). The systems considered in this symposium are summarized in Table 3. Predictions for short-term voyages were good. For long-term voyages and habitats the predictions become fuzzier, centering on algae for food, but failing to consider the problems of raw material conversion and acceptability.

In 1975, a study conducted on space settlements considered agriculture in space using conventional plant culture and animal husbandry (6). Several possible scenarios were considered. In one of these the diet included trout, rabbit, beef, chicken, eggs, and milk, as

Table 3. Anticipated systems for space feeding (5).

Category	System I	System II	System III
Voyage	short	medium	long
Oxygen	carry on	carry on	regenerate
Water	carry on	regenerate	regenerate
Food	carry on	carry on	regenerate
CO_2	chemisorb	chemisorb	use in food production
Waste	store	dehydrate and discard	use in food production

Table 4. Where the food dollar goes (%).

Sector	1971(9)	1977(10)
Farmer	32	31.3
Food Processing & Food Service	39	
Packaging	8	
Transport	5	68.7
Other	16	

(In 1971 column, Food Processing & Food Service, Packaging, Transport, Other bracketed together = 68)

well as a variety of vegetables and cereals. A more realistic scenario might be based on soybeans, rice, wheat, and aquaculture.

Unfortunately, none of the studies so far have seriously considered the problem of conversion of the agricultural raw material into *food*. Yet in the modern society most of the man-hours involved in food production go into this conversion. This is evident from the accounting of costs of food presented in Table 4. Similar analysis of energy expenditures presented in Table 5 again

Table 5. Energy flow in U.S. food chain.

Input of fossil fuel		
	Energy[a]	% of total
Agriculture	7.0	19.7
Livestock operations	2.0	5.6
Food Processing	6.5	18.3
Transportation	4.5	12.7
Sales	5.5	15.5
Food Preparation	10.0	28.2
Total	35.5	100.0
Output of energy in products		
	Energy[a]	% of total
Crop residue and manure	45.0	56
Exports and by-products	6.0	7.5
Human Food	5.0	6.2
Animal metabolism, other waste	24.0	30
Total	80.0	

[a] Joules x 10^{-9}

shows that food processing, distribution, and preparation take the lion's share of energy *(7)*.

Let us consider, then, what is involved in these food conversion operations. Table 6 shows an overall view of food processing operations required to achieve the necessary conversions. Each of these operations is highly specialized. For the achievement of "separation" (step 1c of Table 6), it is possible to use either mechanical or physico-chemical methods. Holdsworth *(8)* lists 19 different methods which may be used for mechanical separation (ranging from "centrifuging" to "washing") and 12 different methods by which physico-chemical separations may be performed. Even for foods considered to be "unprocessed," the required conversions constitute major technological tasks. For instance, Table 7 shows the steps in conversion of rice.

The modern food system is quite complex and possesses certain characteristics which may complicate the adoption of this system for space habitat needs.

Some of the characteristics of modern food systems include:

1. Demand for variety, quality, convenience.
2. Shift in preparation from home to factory.
3. Institutional feeding.
4. Mechanization, automation, large scale of industry.
5. Energy-intensive operations.
6. Use of a variety of composite (engineered) foods.
7. Use of highly specialized processing equipment.

With respect to size, it may be noted that the average single processing plant may cater to the needs of tens of thousands (bakery) or even millions (sugar, oil refineries).

In view of the above considerations, research will be needed before we can even begin to plan a scenario for food technology in space.

The problems in designing the necessary food supply system, and the necessary product and process development, are summarized schematically in Figure 1. This figure represents the major unknowns with respect to acceptability of a hypothetical diet and with respect to the availability of technology to produce the diet. The abscissa represents the types of raw materials which may be produced at a space habitat. They range from simple compounds, such as sugars and amino acids produced by chemical synthesis, to products

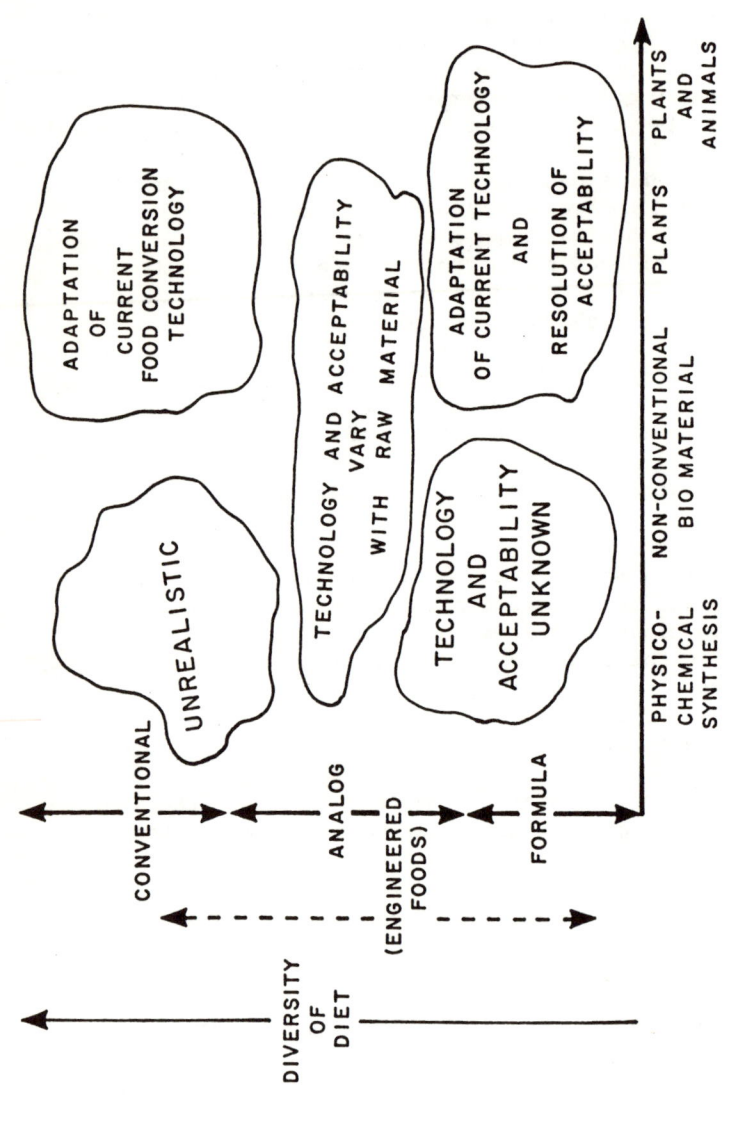

Figure 1. Technological requirements for a range of hypothetical diets, viewed in terms of diet components (x-axis) and for (y-axis).

Table 6. Some food processing operations.

1. Conversion of Raw Materials
    a. Cleaning
    b. Disintegrating
    c. Separation
    d. Heating and Cooling
    e. Mixing
    f. Forming

2. Preservation

3. Packaging

4. Storage and Distribution

5. Preparation for Consumption

6. Waste Disposal

Table 7. Conversion of rice to food.

Rice processing

1. Harvesting (moisture content critical)
2. Drying
3. Cleaning (screens, air aspiration)
4. Dehulling
5. Separation of unshelled rice
6. Milling
7. Brushing
8. Coating
9. Instantizing by heating and redrying

Parboiled rice

1. Harvesting
2. Cleaning
3. Deaeration
4. Soaking
5. Steaming
6. Drying
7. Milling

of conventional agriculture encompassing plants and animals. The ordinate represents the variety and quality of the diet. It ranges from a nutritionally complete but organoleptically far from satisfying, "formula food" (perhaps a combination of liquid formula and wafers) to the conventional "steak, brocoli and ice cream", or "hamburger, potatoes and fresh fruit" diets. For the sake of simplifying the analysis of the problems attendant with the selection of the raw material/diet options, we have indicated five regions in the diagram. The possibility of producing conventional diets from chemically synthesized raw materials appears at present as an unrealistic one. The production of formula diets from conventional raw materials presents no major technological problems, if the raw materials can be produced. There is a large unknown, however, in the acceptability of such formula diets. Furthermore, production of diets from chemically synthesized or non-conventionally produced ingredients (synthetic chemicals, bacterial proteins, plant cells grown in culture, etc.) is at present only an untested scheme. Much research on both the technology and the acceptability aspects will be required. The production of a conventional diet from conventional raw materials, but under conditions of space habitats, appears a feasible proposition, once the agricultural materials can be produced. Much engineering research and development will, however, be required before processing technology appropriate to space can be produced. Some of these problems will be discussed later.

A possible approach will be the production of engineered foods. These foods may be analogs of conventional foods (e.g. cheeses made of soy instead of cow's milk; or meat-like chunks made of gluten and other vegetable proteins) or they may be entirely new foods compounded from available ingredients and made to conform to the requirements with respect to nutrition, acceptability, and stability. Since they may include a variety of ingredients, and since they may be produced in equipment capable of serving a variety of foods (not specialized for only one type of ingredient), they may be ideal for utilization in space.

Before any of the potential scenarios can actually be considered in adequate detail, and certainly before any of these scenarios can be tested by large-scale experiments, much research will be needed.

Table 8 summarizes some of the research needs in the area of nutrition. Several specific nutrient requirements--

Table 8. Human nutrition in an isolated environment - Research needs.

1. Requirements for trace nutrients

2. Special requirements of the habitat

    a. Radioprotective agents of the habitat
    b. Metabolism changes in new conditions of gravity and pressure
    c. Accumulation of trace toxins in total recycling
    d. Capability of limited menus to provide not only "adequate", but "optimal" nutrition

3. Tailoring nutrition to specific population group needs

4. Nutrient content and stability of foods grown and processed under habitat conditions

5. Role of nutrition in maintaining health in the habitat

---

especially with respect to trace minerals--are still unknown. In addition, the effect of specific space conditions on nutrient requirements must be established. Another area requiring research is the acceptability of various types of diet under space conditions. The variety required, the type of ingredients that would be tolerated psychologically and organoleptically, and other cultural aspects of feeding will require further study.

As was indicated previously, the process and equipment design for food production and preparation under space conditions will require development. Table 9 shows some of the problems inherent in this development and the associated research tasks.

Other major areas for research include the impact of the food technology in the habitat on its ecology, consideration of various capabilities for resupply in either normal or emergency situations, and, last but not least, the social and human factors. Some of these are summarized in Table 10.

As we can see, considering the technology of food supply in space leads one to a variety of research topics. There is no doubt in my mind that sooner or later these research tasks will be undertaken, solutions found, and man supported in his endeavor to explore space. The benefits of this research to food technology on earth will

Table 9. Research and development tasks for food technology in space.

Process and Equipment Design
1. Flexibility and versatility
2. Small scale of operations

3. Adaption to habitat conditions
    a. Lack of chemical and noise pollution
    b. Reduced atmospheric pressure operation
    c. Different "gravity" conditions
    d. Utilization of solar energy and "hard vacuum"
    e. Provision for "total recycling"

4. Adaption to "remoteness" from earth industries
    a. Maintenance and replacement of parts
    b. Fail-safe operations
    c. Simplicity
    d. Minimize utilization of chemicals

5. Provide capability for modification: Provide for "the unexpected"

Table 10. Research and development tasks for food technology in space.

Social and Human Factors
1. Kitchen, vending machine, cafeteria?
2. Social and ethical aspects of medication delivered through food and water supply (radiation protection, tranquilizers, antibiotics, fluoride, birth control agents?).
3. Legal and enforcement aspects. (Will chewing gum, chocolate bars, and coffee become black market items? Food and Drug Laws and enforcement for habitat.)

be immense. Many of the problems of feeding mankind under a variety of restraints will be solved by the research necessary for solving problems of man's existence in space.

References and Notes

1. M. Modell, "Sustaining Life in a Space Colony," *Tech. Rev.* (July/Aug. 1977), p. 36.

2. J. Verne, *From the Earth to the Moon*, Translated by E. Roth (Dover Publ., New York, 1960).

3. M.C. Smith, N.D. Heidelbaugh, P.C. Rambaut, R.M. Rapp, O.H. Wheeler, C.S. Huber, C.T. Bourland, Apollo Food Technology, In *Biomedical Results of Apollo*, R.S. Johnston, L.F. Dietlein, C.A. Berry, Eds. (NASA SP-368, National Aeronautics and Space Administration, Washington, D.C., 1975).

4. J. Verne, *To the Sun and Off on a Comet*, Translated by E. Roth (Dover Publ., New York, 1960).

5. M.C. Brockmann, A.C. Henick, G.W. Kurtz, R.G. Tischer, "Closed Cycle Biological Systems for Space Feeding," *Food Technol.* *12* 9 (1958), p. 449.

6. R.D. Johnson and C. Holbrow, Eds., *Space Settlements: A Design Study* (NASA SP-413, National Aeronautics and Space Administration, Washington, D.C., 1977).

7. F.A.O. (Food and agriculture organization of the United Nations), *The State of Food and Agriculture* (United Nations, New York, 1976), p. 106.

8. S.D. Holdsworth, "Chemical Engineering and Food Preservation Research, Trends, and Reflections," *Chem. Ind.* (1971), p. 16.

9. A.L. Brody, "Packaging's Indispensable Role in the Food Industry," *Food Technol.* *28* 8 (1974), p. 35.

10. U.S. Department of Agriculture, *Developments in Marketing Spreads for Food Products in 1977* (AER No. 398, Economics, Statistics, and Cooperative Service, Washington, D.C., 1978).

C. David Raper, Jr., Terry A. Pollock, Judith Fey Thomas

9. Use of Phytotrons in Assessing Environmental Requirements for Plants in Space Habitats

Introduction

The process of photosynthesis is central to the inclusion of plants in a Closed-Environment life support system (CELSS) for space habitats of a mission duration and crew size that render food and atmospheric regeneration desirable. Within the chloroplasts of green plants the energy from photosynthetically active radiance (PAR) of wavelengths between 400 and 700 nm is utilized for "photolysis" of water to hydrogen and hydrozyl ions. The reducing potential of the hydrogen ions is utilized to reduce carbon dioxide to the level of simple sugars, while the oxidizing potential of the hydroxyl ions is utilized to evolve oxygen. These biochemical processes of photosynthesis are summarized in the following reactions:

$$4 H_2O \xrightarrow{PAR} 4(H^+) + 4(OH^-) \qquad (1)$$

$$4(H^+) + CO_2 \longrightarrow (CH_2O) + H_2O \qquad (2)$$

$$4(OH^-) \longrightarrow 2H_2O + O_2 \qquad (3)$$

Thus, the products of the photosynthetic process are: (1) the basic carbon units of carbohydrates from which the amino acids, vitamins, carbohydrates, and other organic components essential for human nutrition are synthesized by complex plant and animal metabolism and (2) the regeneration of oxygen required for respiration. The reduction of carbon dioxide concomitant with oxygen evolution and carbohydrate regeneration is required for an essentially self-sustaining, regenerative CELSS. Although reduction of carbon dioxide with evolution of oxygen can be accomplished either biologically by photosynthesis or with physico-chemical systems, only the biological photosynthetic process is capable of reincorporating carbon into the organic constituents of human food.

Higher plants are a complex and environmentally responsive system. For plant agriculture to reliably and predictably meet the requirements for food production and oxygen evolution in a CELSS, tolerable and optimal environmental limits for plant growth must be established experimentally. *Phytotrons are laboratories designed for the study of plant responses to environment, and are defined as a collection of controlled-environment cabinets, rooms, and glasshouses organized so that many combinations of independently variable environmental factors can be studied simultaneously (1).* There are presently three phytotrons in the United States, on the campuses of North Carolina State University, Duke University, and the University of Wisconsin. Although the plant-growth chambers in these facilities are open systems, plants can be grown with precisely controlled variations in temperature, humidity, nutrition, radiance, relative humidity, and carbon dioxide (Table 1).

Two workshops have been organized at Ames Research Center by the Office of Life Sciences of the National

Table 1. Control limits in the North Carolina State University Phytotron (1).

Temperature:	aerial	$-10°$ to $40°$ (± 0.25) C
	root	$14°$ to $35°$ (± 0.50) C
Thermoperiod:		any diurnal schedule of day and night temperatures
Radiance:	PAR	0 to 75 (± 2.5) nEinsteins $cm^{-2} sec^{-1}$
	PR	0 to 12 (± 1.0) w m^{-2}
Photoperiod:		any diurnal schedule of light and dark
Relative humidity:		40% to 95% ($\pm 5\%$) at $14°$ to $34°C$
Carbon dioxide:		150 to 2000 (± 25) 1/liter
Nutrition:		No limits

Aeronautics and Space Administration and the Bioenvironmental Systems Study Group, a team coordinated by the Society of Automotive Engineers, Inc., to assess the state-of-the-art for inclusion of higher plant agriculture in CELSS. Recommendations for research and development made by these workshops include three broad areas which can be investigated in existing phytotron facilities: (1) a systematic determination of detailed changes in plant morphology and chemical composition in response to variations in individually controlled environmental factors; (2) use of these data to develop a dynamic crop simulation model with multiple crop capacity and feed-forward interactive capabilities for optimizing the plant environment; and (3) evaluation of cultural and genetic manipulation to idealize plant phenotypes for space habitats.

Changes in Plant Growth Attributable to Environment

Plants are being grown commercially and experimentally in controlled environments in both natural and artificial light. Thus, it can be assumed that the minimum technology for intensive plant culture in controlled-environment agriculture already exists. However, there is little consistency in growth and yield. If plant agriculture is to supply the nutritional and the carbon dioxide and oxygen recycling needs of a CELSS, growth and production must be reliable and the entire range in responses of plant performance to environmental selection must be identified.

Phytotrons provide facilities for immediate investigations into environmental effects on plant growth, morphology, and physiology. An example of the importance of selection of environment on the efficiency of plants in producing edible biomass and recycling oxygen and carbon dioxide is the interactive effects of photoperiod and temperature during pod development on growth and yield of soybeans (Table 2). Total biomass produced and average total growth rate increased with temperature, and at all temperatures were greater in the long-day than short-day photoperiods. The photoperiod duration, which was varied between treatments only after anthesis so that floral initiation and pod set were unaffected, was extended to long-days with photomorphogenic radiation (PR) of wavelengths between 700 and 850 nm. Since the chlorophyll pigment involved in photosynthesis does not respond to PR, the increase in rates of total biomass production under long-days can be attributed to enhanced efficiency of utilization of the incident PAR. The effects of temperature and photoperiod on edible biomass were interacted. Edible biomass increased with temperature under short-days but decreased with temperature under

Table 2. Effects of photoperiod and temperature during pod-fill on growth of soybeans (2).

Photoperiod	Day/Night Temperature (°C)	Biomass		Efficiency for Edible Harvest (%)	Growth period (days)	Average rate of Biomass Production (gm^{-2}day^{-1})
		Total (gm^{-3})	Edible (gm^{-2})			
Short-day	30/26	2225	968	44	114	20
Long-day	30/26	4375	355	8	132	33
Short-day	26/22	2097	903	43	116	18
Long-day	26/22	3677	419	11	141	26
Short-day	22/18	1634	624	38	121	14
Long-day	22/18	3215	1000	31	139	23

long-days, so that equivalent seed yields were obtained under long-days at 22/18 °C and under short-days at 30/26 °C. At 22/18 °C, the higher yield of edible biomass under long-days was achieved with little change in the ratio of edible to total biomass (or efficiency of edible harvest). The amount of oxygen regeneration and carbon dioxide removal are directly related to the rate of total biomass production; thus, with manipulation of temperature and photoperiod, systems engineers are given a range of options for efficiencies of oxygen recycling and food production by a soybean crop, depending on the temporal requirements for balancing the total functions of a space habitat.

This example of manipulation of plant functions by environmental selection is especially pertinent to a discussion of the role of phytotrons in research and development for CELSS. In the natural field environment for soybeans, photoperiod and temperature both decrease during pod development. Until this phytotron study, the efficiency of edible harvest, which is affected only slightly by temperature, appeared to be a fixed genetic characteristic of soybeans. But this is only a single example of the potential responsiveness of crop plants to environmental selection. In order to develop scenarios for combinations of crop species and performance levels to optimize the functions within a CELSS, extensive research in phytotrons will be required.

If there is scant knowledge of environmental influence on edible biomass productivity of crop plants, there is even less quantitative information on the effects on the nutritional quality of the edible plant portions. Most agricultural research has been directed toward increasing the mass of food production, with confidence in the presumption that the diverse crop sources and geographic origins of the raw plant materials in an earth diet minimizes the likelihood of an environmentally related nutritional deficiency in the composite diet. In a space habitat, with possibly limited plant species and cultural systems, there can be no presumption of nutritional balance through composite food sources. The effects of nutritional composition of edible biomass become as critical as the quantity of edible biomass.

Although it is sparse and qualitative, there is evidence of the environmental effects on nutritional value of crop yield. In example, among seed such as soybeans, high in both oil and protein, increased availability of nitrogen enhances the protein content and reduces the oil content (Krober, 1956). A moisture stress during soybean seed development increases oil composition of the beans at the expense of protein (3). Although oil composition of soybeans is also increased by higher temperature, there the total composition of protein is not affected. However, the composition of methionine, the essential amino acid present in soy protein in the most limiting quantities for a balanced protein source, is decreased as temperature is lowered (4). This reduces even more the nutritional quality of soybean as a source of complete protein. There are also other potential nutritional consequences in the manipulation of environment to increase plant productivity. Associated with the enhancement of plant growth by carbon-dioxide enrichment of the atmosphere is a reduction in the nitrate composition of leaves (5). Low levels of nitrate, a potential toxin and carcinogen, is a dietary advantage. However, other potential toxic elements, such as manganese, when present in water in moderate concentrations, can be accumulated in abnormally high concentrations in plants grown at high temperatures (6).

Dynamic Crop Simulators

The finite pools of essential materials causes a CELSS to be a minimally buffered system. Because many of these materials, including oxygen, water, mineral nutrients, and carbon dioxide would be cycled and regenerated through the plant agricultural system, an accurate projection is necessary of the amounts of these limited materials tied up in standing and stored plant materials at any time, and of the rates of incorporation into a release from the plants at any time. These factors vary with species and stage of growth of plants. They also vary with environment. Obviously, a deterministic dynamic simulation model must be developed for the plant agricultural system. For a computer simulator capable of the accuracy and precision required by an inflexible CELSS, physiological and morphological responses to all combinations of environmental variables must be documented. Phytotrons are well suited to facilitate experimental studies required to provide the data base for construction of comprehensive crop simulators. Because all environmental factors can be controlled independently, phytotron experimentation facilitates accurate information on the physical and chemical environment and mathematical descriptions of rates of plant processes and growth as functions of specific changes in environment. Most crop models being developed for current agricultural systems depend on phytotron studies for physiological and ontogenetical detail (7,8).

A research undertaking to experiment with the effects and interactions on growth and development among even a normal range for all environmental parameters required for comprehensive model construction is an immense task. Fortunately, as a mechanistic simulation model is developed from controlled-environment data, it becomes effective in truncating the combinations of environmental variables required to document plant-environmental interactions. The number of experimental environment-plant combinations is reduced to the few levels necessary to establish and verify a response surface, especially as the model is expanded to include new crop species. Thus, the model becomes a tool for increasing the efficiency of phytotron research.

As the sensitivity of the crop simulator is increased by experimentation, it can become a feed-forward interactive system for controlling the plant environment. Coupled with environmental and plant monitoring systems, the crop simulator can project final mass or anticipated oxygen and carbon dioxide cycling rates from attained growth and suggest down-the-line environmental alternatives to optimize plant functions so as to maintain a balance in the CELSS. The fundamental logic for such dynamic crop simulation systems is already being developed utilizing phytotron experimentation.

Cultural and Genetic Manipulations

Crop plants possess a certain amount of genetic plasticity that potentially can be used to breed plants with a more favorable ratio of edible to total biomass or a nutritionally superior composition of plant products. However, because of interactions between genetic and environmental factors there is a high probability that genetic effects may be missed or incorrectly assessed unless environmental effects are taken into account (9). Previously discussed (Table 2) has been an example of environmental factors (photoperiod duration and temperature) which alter the ratio of edible to total biomass of soybeans over a range of 8% to 44%. Other factors such as partial pressure of oxygen can shift the ratio of edible to total biomass over a similar range (10). The controlled-environment capabilities of phytotrons provide identifiable and stable environments which can enhance efforts to selectively breed for idealized phenotypes.

Environment not only affects growth and development of the current generation, but it can also—through changes in mother-plant and seed development—alter growth responses of subsequent generations (11). The stability of progeny performance must be investigated throughout several generations within controlled environments.

Other Factors

Phytotrons are being used to evaluate plant response to man-made pollutants and toxic elements. The sensitivity of plant response to gaseous and trace element toxins is greatly influenced by the presence of other environmental stresses (12). Much additional research remains to be done on the interactive effects of environmental and man-made phytotoxins on growth. An area only beginning to be explored is the potential phytotoxic effects of compounds released into the environment by the plants themselves and by materials used in construction of plant culture systems (12). In open systems, these compounds may be diluted to a sufficient extent that phytotoxic symptoms are avoided. In closed systems, these compounds can become critical factors in plant growth. Also to be considered is that many of the potentially phytotoxic materials released by plants are secondary metabolic compounds and growth regulators which are produced in greater quantity when the plants are under stress. Hence, not only does environment alter the sensitivity of plants to toxins, it can stimulate the plant to introduce additional toxins into its environment.

Summary

A brief overview has been given in this paper of some of the uses of phytotrons in assessing the environmental requirements for plants in space habitats. The discussion has been directed toward attainment of that knowledge of plant growth responses to environment which is essential before the successful inclusion of plant agriculture in a CELSS can be assured. This same knowledge is necessary in the approaching decades if earth agriculture is to continue to meet the expanding requirements for plant products by an expanding population, in the face of limiting resources. The existing agricultural research technology represented by phytotron facilities can be used in providing answers to questions of the future of plant agriculture in space habitats and on earth.

References and Notes

1. R.J. Downs and V.P. Bonaminio, "Phytotron Procedural Manual for Controlled-Environment Research at the Southeastern Plant Environment Laboratories," *North Carolina Agricultural Experiment Station Technical Bulletin No. 244* (1976), pp. 1-36.

2. C.D. Raper, Jr., and J.F. Thomas, "Photoperiodic Alteration of Dry Matter Partitioning and Seed Yield in Soybeans," *Crop Science 18* 5 (1978), pp. 654-656.

3. N. Sionit and P.J. Kramer, "Effect of Water Stress During Different Stages of Growth of Soybean," *Agronomy Journal 69* 2 (1977), pp. 274-278.

4. O.A. Krober, "Methionine Content of Soybeans as Influenced by Location and Season," *J. Agric. Food Chem. 4* 3 (1956), pp. 254-257.

5. C.D. Raper, Jr., W.W. Weeks, R.J. Downs, W.H. Johnson, "Chemical Properties of Tobacco Leaves as Affected by Carbon Dioxide Depletion and Light Intensity," *Agronomy Journal 65* 6 (1973), pp. 988-992.

6. W.T. Rufty, G.S. Miner, C.D. Raper, Jr., "Temperature Effects on Growth and Manganese Tolerance in Tobacco," *Agronomy Journal 71* 5 (1979), in press.

7. J.D. Hesketh, H.C. Lane, J.W. Jones, J.M. McKinion, D.N. Baker, A.C. Thompson, R.F. Colwick, "The Role of Phytotrons in Constructing Plant Growth Models," *Phytotronics in Agricultural and Horticultural Research. Phytotronics III*, (Gauthier-Villars, Paris, 1976), pp. 117-129.

8. M. Wann, C.D. Raper, Jr., H.L. Lucas, Jr., "A Dynamic Model for Plant Growth: A Simulation of Dry Matter Accumulation for Tobacco," *Photosynthetica 12* 2 (1978), pp. 121-136.

9. V.V. Rendig and D.S. Mikkelsen, "Plant Protein Composition as Influenced by Environment and Cultural Practices," In *Opportunities to Improve Protein Quality and Quantity for Human Food*, Special Publication No. 3058 (University of California, Davis, 1976), pp. 84-106.

10. B. Quebedeaux and R.W.F. Hardy, "Oxygen Concentration: Regulation of Crop Growth and Productivity," In

CO_2 *Metabolism and Plant Productivity* (University Park Press, Baltimore, 1976), pp. 185-204.

11. J.F. Thomas and C.D. Raper, Jr., "Germinability of Tobacco Seed as Affected by Culture of the Mother Plant," *Agronomy Journal* 71 4 (1979), in press.

12. D.P. Ormrod and D.T. Krizek, "Environmental Stresses in Controlled Environments," *Phytotronic Newsletter 19* (Secretariat Phytotronique, Gif-sur-Yvette, France, 1978), pp. 41-51.

_____ John M. Phillips

10. Controlled-Environment Agricultural Systems for Large Space Habitats

Background

The future space program of the United States is linked to the near-term deployment of the Space Transportation System (STS), or Space Shuttle. The Space Shuttle will provide routine access to the space environment at a cost that will be greatly reduced over that of previous booster systems. The STS features reliance on reusable components to cut transportation costs. Once the Shuttle is successfully deployed, a series of scientific and technical missions will be flown which will greatly expand the role of space technology in creating a positive future for all humanity. Shuttle missions will be concerned with resource monitoring, pollution abatement, crop, ocean and forestry surveying, communications, improved weather forecasting, space processing, life sciences, and a host of other basic and applied investigations which will greatly advance the frontiers of human knowledge (1).

During the 1980s, increasingly ambitious projects will be attempted in space, leading evolutionarily to the development of large space structures such as communication arrays and solar satellite power systems (SSPS) (2). These ventures, in turn, will require the development of large space habitats to house the construction and maintenance crews that will build and operate such systems. Numerous studies have been conducted in recent years which suggest that the vast material and energy resources of space could be utilized to help solve the many problems confronting our species as we approach the end of this millenium (3,4,5). By the year 2000 AD, humanity will have to evolve solutions to shortages of energy, material resources, food, and space to accommodate an ever-increasing population and the burgeoning expectations for improved life styles. In many respects, the expansion of human activity to the extraterrestrial environment of the solar system represents an option-increasing enterprise which can be anticipated to return manifold benefits to society. Such benefits may include renewable energy supplies from space, space-processed materials and products, and most prominently, a wealth of new technology and knowledge which will allow us to better cope with the demands of the future (6,7).

One of the technologies which will have to be developed to sustain an expanded role for human activity in space is closed-environment life support systems (CELSS) for large space habitats. CELSS will be required to support the crews of space manufacturing facilities and other large-scale ventures involving substantial numbers of people at sites that are rather remote from earth. Previous, small-scale space missions could rely on stored foods and other consumables (such as gas absorption materials) because of the limited numbers of personnel involved, the short mission duration, and the convenience of resupply. However, for large-scale space missions, transport costs to supply fresh foods to the crew of a space manufacturing facility or similar mission provide a powerful driver for the development of life support systems that are essentially "closed-cycle" with respect to materials. In these systems, waste products would be recycled into inputs by various technologies which will most probably involve a combination of biological and physico-chemical processes. In general, life support systems will approach closure of material recycling due to food regeneration *in situ*, as food production provides the logical method of recycling waste products of human activity into usable inputs.

There are many problems to be encountered in the design and development of food production systems for large space habitats. Diet scenarios must be selected which are suited to the needs of the crew. Crop species must be chosen and suitable cultural systems must be developed. The use of animals in the system must be given careful scrutiny because of the possibility of cross-contamination from some pathogens which are shared by both humans and animals (8). The system must be safe. It must be characterized by reliability and predictability, and it must be relatively simple and labor-efficient to maintain. These and many other characteristics of CELSS have been reported in the numerous recently conducted studies that have focused on this problem (9,10,11,12). We would like to discuss here the potential role of a terrestrial technology which may prove to be highly relevant to the problem of developing agricultural systems for large space habitats. This technology is known as controlled-environment agriculture (13).

Controlled-Environment Agriculture

Controlled-environment agriculture (CEA) is a food production technology under development and expansion in the United States, Western Europe, the Soviet Union, Japan, and elsewhere *(14)*. CEA is characterized by food production activity conducted within an enclosure wherein crop production factors are readily subjected to human manipulation *(15)*. The use of greenhouses for winter cultivation of fresh vegetables such as tomatoes is a typical example of CEA. Commercial acreages of CEA facilities are largest in the highly industrialized societies where suitable land and climate are not available for year-round production of certain high-demand crops. Thus, the greenhouse industry in the United States is relatively insignificant in size due to the availability of outdoor production areas with mild winter growing conditions. In contrast, substantial acreages are found in Europe and Japan where land is limited and climate unfavorable. In the Soviet Union, unfavorable climate is the primary motivator for the production of tree crops, in addition to the more typical vegetables, in greenhouses.

CEA systems are under development for animal food species, and include confinement rearing systems for poultry, swine, and other livestock, and controlled-environment aquaculture systems for the production of shrimp, prawns, fin fish, and other aquatic organisms.

The use of an enclosure removes the influence on production of such external factors as climate and weather and allows the internal environment to be optimized for maximum yield. Year-round production becomes feasible and the benign microclimate of the enclosure leads to accelerated growth rates. Yields of vegetable crops in CEA are dramatically increased over those experienced in open-field farming systems *(16)*. CEA systems confer numerous other benefits to food production activity besides increased yields. In controlled systems, crop production becomes highly predictable. This allows the scheduling of crops to meet an anticipated need or desired production goal. The use of CEA also leads to efficient resource utilization. Crop growth inputs, such as water, nutrients, and other factors, are confined to the system, thereby eliminating leaching and other effects which limit resource-use efficiency in open field-systems.

A negative aspect of CEA systems is that they are capital-, energy-, labor-, and material resource-intensive *(17)*. Production costs for food grown in CEA exceed those

of open-field systems. However, it must be realized that
CEA systems have been developed precisely to fulfill the
need for crop production under conditions where open-field
farming systems are impracticable or unworkable due to
climate, weather, or other reasons.

Thus far, basic food crops such as the cereal grains
and legumes have not been produced on a commercial scale in
CEA due to the unfavorable economics. Although yields of
these crops in CEA facilities could be ten to twenty times
that experienced in open-field systems, economic returns
would not be sufficient to offset the high costs of production at present and projected prices for such basic
commodities. Studies have been conducted with basic food
crops in phytotrons* and other CEA research facilities *(18)*.
These have been mainly concerned with plant physiology.
Only a few experiments have sought to evaluate the influence
on yield of edible products of manipulating the growing
environment within an enclosure *(19,20)*. Additional studies
are required before we can fully evaluate the potential for
CEA for cereals and other basic foods.

In many respects, the need to develop food production
systems for large space habitats will provide a mandate to
undertake studies of basic food crops in CEA-type systems.
This research effort may also be viewed as contributing to
our food production strategy for the future. For example,
CEA production of basic food crops may become economically
viable due to a number of factors which could become dominant food production considerations in the future. These
include changes in climate, such as the onset of a new ice
age or more variable weather patterns than those which have
prevailed for the last 50 years or so, contamination of the
world food supply by the buildup of pesticide residues,
heavy metals, or other toxins, and on the more positive
side, dramatically increasing demand due to rising expectations and available income in the centrally planned and
developing countries.

CEA production of basic food crops will be preceded, of
course, by a large expansion of acreage in the more traditional, high-value crops such as vegetables and melons.
Already, in Japan, a cropping pattern has developed where
low-cost plastic greenhouses are erected on rice-growing
land during the winter fallow period for the production of

---

* See C. David Raper, Jr., Terry A. Pollock,
Judith F. Thomas, "Use of Phytotrons in Assessing
Environmental Requirements for Plants in Space Habitats,"
elsewhere in this publication.

melons *(21)*. We may expect that, at some point, the
Japanese will have developed CEA systems which are low
enough in cost to justify their use on the rice crop as well
as for the production of the high-value melon crop.

The technology of controlled-environment food production is obviously relevant to the problem of developing
controlled-environment agricultural systems for large space
habitats. Terrestrial CEA represents a data base of experience and research which has brought about viable food
production systems well-suited to the demands of producing
food in abundance in a hostile environment. Much of the
technology already developed for CEA could be adapted with
minor modifications to the problems to be encountered in
developing a food production system for large space habitats.
Additional research effort invested in the technology of
controlled-environment food production will yield information that will benefit both space and earthly applications.
The potential technological returns of the research effort
to develop controlled-environment food production systems
are likely to be far-reaching in their impacts and benefits
to society in general.

Future Research

A long-term research and technology development effort
must be mounted in order to create flight-ready food
production systems for large space habitats. Much of the
required research can be performed in ground-based
facilities such as phytotrons and research greenhouses. A
series of fundamental investigations will have to be
conducted in space to evaluate the influence of the space
environment on potential biological components of CELSS.
These investigations must determine if other than earth-normal design criteria, such as partial gravity, would be
acceptable for use in the food production systems of large
space habitats. If earth-normal design criteria must be
adopted, this will materially affect the mass requirements
and costs for the habitat.

Near-term and short-term research requirements for the
development of food production systems for CELSS have
recently been reported in detail *(22)*. Major objectives of
this suggested research program include the following:

1. Crop and animal species selection.

2. Development of resource-efficient production
   systems.

3. Quantification of inputs and outputs, including the production of gaseous by-products and other contaminants.

4. Evaluation of various waste-recycling systems including biological and physico-chemical processes.

5. Development of monitoring and control systems.

6. Systems engineering and integration into a functioning food production unit which is compatible with other life support system components.

7. Spaceflight evaluation of the influence of the space environment on biological components and any subsystems involving gas/liquid phase changes.

A major milestone in the proposed research program will be the development of a ground-based demonstration and testing facility which can be operated as a manned, closed-chamber life support system. This facility must be large enough to provide sufficient information to allow scaling-up and expansion of the system to accommodate crew sizes for anticipated space missions. Since most of the research effort necessary for the development of controlled-environment agricultural systems for large space habitats will be performed under ground conditions, immediate tangible returns in the form of terrestrial impacts and benefits should be realized.

Terrestrial Applications and Benefits

Throughout this paper we have alluded to the wealth of knowledge and other spin-offs which will be generated from the research and technology development effort required to create controlled-environment agricultural systems for large space habitats. Some of the terrestrial applications and benefits to be derived from this research program are discussed below:

1. A functioning, *in situ* food production facility designed to feed all the inhabitants of a space settlement and industrial complex will make the concept truly beneficial to earth. There will be no constant drain of materials away from earth to support such sites. With life support an autonomous function, the base can even provide a genuine first outpost on the road to human population expansion away from earth. Without a self-contained agricultural component, however, the space manufacturing site would be just another drain on the food-producing resources of the planet.

2. One of the most important benefits of developing a CEA food production system for space is that there will be immediate feedback--indeed, a great deal of interplay-- between space and earthly CEA systems. Earth systems are already significantly contributing to the food supplies of congested areas like Europe and Japan and to areas with severe weather problems like Russia and Abu Dhabi. As people continue to build on agricultural land and also allow such land to be eroded away or lost to cultivation through salt accumulation, the necessity for more efficient CEA systems will become apparent even in land-rich places such as the United States. As the puzzles of how to save space, time, weight, materials, and energy in space systems are solved, better and more economical greenhouses will become available through the commercial spinoffs. This is the kind of technology that has high initial costs for research and development, but which is readily adapted to cheap proliferation after the first models are completed.

3. Since full recycling will be necessary for large-scale, long-term food production in space, a great deal will have to be learned about efficient use of various products and by-products. This knowledge will undoubtedly be of use in solving the problems of efficient recycling on earth. Particularly important will be the computer linkages that will monitor and automatically regulate various flows of materials. With good, complex recycling systems available, a possible proving ground for terrestrial applications would be the inner cities. Here there is the space and the need to try new technology for relatively small areas where resources of all kinds have been depleted to the point where a comfortable life cannot be maintained. With such new technology available, there may be no need to repeat the mistakes of environmental degradation when seeking to create or expand urban environments.

4. Life support development research will provide information on the quantitative requirements to sustain human life. This knowledge has applications in many fields, particularly with social programs whose successes and failures often turn on points of physical need which are unnoticed or unknown.

5. Productivity research in agriculture, widely acknowledged to be a prime research need on which the government has not yet moved with sufficient vigor, should receive a substantial impetus from this new research program.

6. At the base of economic health is a nation's productivity, and this depends directly on individual humans.

If they feel good about what they are doing, they work better and GNP increases. In order for people to feel like working, they have to have a dream. They have to believe that, by their own efforts, horizons can be expanded and their situation in life can be improved. Without this hope, productivity declines. Space is a big enough dream, but it must be made accessible to the individual. Space research concerned with food production touches on a subject that is within the common experience of everyone. Research on controlled-environment food production systems for large space habitats may do much to rekindle society's interest in support for space research in general, since food is a universally familiar subject. A renewed interest in space and the future possibilities for the human race could do much to promote a more positive attitude on the part of individuals in a society that has come to regard the future with uncertainty, and even with a sense of hopelessness.

Conclusions

In order to expand the role of humans in space to bring benefits to earth, life support systems for large space habitats must be developed which include controlled-environment agricultural systems to produce food from recycled wastes. The research program required to develop this technology can be performed largely in ground-based facilities and will yield considerable knowledge with terrestrial applications and impacts. A terrestrial technology known as controlled-environment agriculture may contribute a great deal of information that would be useful in the development of food production systems for large space habitats.

References and Notes

1. National Aeronautics and Space Administration, Lyndon B. Johnson Space Center, *Space Shuttle* (Scientific and Technical Information Office, NASA, Washington, D.C., 1976).

2. J. Von Puttkamer, Developing Space Occupancy: Perspectives of NASA Future Space Program Planning, In *Space Manufacturing Facilities II (Space Colonies)*, J. Grey, Ed. (American Institute of Aeronautics and Astronautics, New York, 1977).

3. R.D. Johnson and C. Holbrow, Eds., *Space Settlements: A Design Study* (NASA SP-413, Scientific and Technical Information Office, NASA, Washington, D.C., 1977).

4. K.A. Ehricke, *Industries in Space to Benefit Mankind: A View Over the Next 30 Years* (NASA Contract Report NAS8-32198, Rockwell International Corp., Downey, California, 1978).

5. Science Applications, Inc., *Space Industrialization Study: Part 1 - Final Briefing* (Item 740-049/369 U. S. Government Printing Office, Washington, D.C., 1977).

6. J. Grey, Ed., *Space Manufacturing Facilities I (Space Colonies)* (American Institute of Aeronautics and Astronautics, New York, 1977).

7. J. Grey, Ed., *Space Manufacturing Facilities II (Space Colonies)* (American Institute of Aeronautics and Astronautics, New York, 1977).

8. J. Phillips, et al., *Studies of Potential Biological Components of Closed Life Support Systems for Large Space Habitats: Research and Technology Development Requirements, Costs, Priorities, and Terrestrial Impacts* (Final Report NASA Grant NSG-2309, Advanced Life Support Projects Office, NASA Ames Research Center, Moffett Field, California, 1978).

9. H.K. Henson and C.M. Henson, Closed Ecosystems of High Agricultural Yield, In *Space Manufacutirng Facilities I (Space Colonies)*, J. Grey, Ed. (American Institute of Aeronautics and Astronautics, New York, 1977).

10. R.D. Johnson and C. Holbrow, Eds., *op. cit.*

11. J.M. Phillips, Controlled-Environment Agriculture and Food Production Systems for Space Manufacturing Facilities, In *Space Manufacturing Facilities II (Space Colonies)*, J. Grey, Ed. (American Institute of Aeronautics and Astronautics, New York, 1977).

12. J.M. Spurlock and M. Modell, *Technology Requirements and Planning Criteria for Closed Life Support Systems for Manned Space Missions* (NASA Contract NASw-2981, Office of Life Sciences, NASA, Washington, D.C., 1978).

13. J.M. Phillips, A.D. Harlan, K.C. Krumhar, "Developing Closed Life Support Systems for Large Space Habitats," Paper AAS-78-145 presented at the 25th Anniversary Conference of the American Astronautical Society, Houston, Texas, October 30, 1978.

14. D.G. Dalrymple, Controlled Environment Agriculture: A Global Review of Greenhouse Food Production (Foreign Ag. Econ. Rpt. No. 89, Economic Research Service, U. S. Department of Agriculture, Washington, D.C., 1973).

15. International Research and Technology Corporation, Inc., *An Assessment of Controlled Environment Agriculture Technology* (IRT-469-R-Final, National Science Foundation, Washington, D.C. 1978).

16. M.R. Fontes, "Controlled-Environment Horticulture in the Arabian Desert at Abu Dhabi," *Hortscience 8* 1 (1973), pp. 13-16.

17. International Research and Technology Corporation, Inc., *op. cit.*

18. R.J. Downs and H. Hellmers, *Environment and the Experimental Control of Plant Growth* (Academic Press, New York, 1975).

19. S. Yoshida, "Effects of Carbon Dioxide Enrichment at Different Stages of Panicle Development on Yield Components and Yield of Rice," *Soil Sci. and Plant Nut. 19* 4 (1973), pp. 311-316.

20. J.J. Riley and C.N. Hodges, "Plant Response to Carbon Dioxide Enrichment: A Function of the Microenvironment," Paper presented at the meeting of the Southwestern and Rocky Mountain Division, AAAS, Colorado Springs, Colorado, May 7-10, 1969.

21. D.G. Dalrymple, *op. cit.*

22. J. Phillips, *et al., op. cit.*

_____ B. P. Miller

11. Economic Factors of Outer Space Production

Introduction

The purpose of this paper is to examine economic and policy issues related to the use of earth-orbiting satellites to produce goods and services vital to mankind. The nature of the goods and services that could be provided by earth-orbiting satellites will be reviewed, along with economic issues such as costs, benefits, and the market for the goods and services. The interaction between the public and private sectors in the development and ownership of the systems which could produce the goods and services will also be discussed.

The Nature of Outer Space Production

In defining the nature of the goods and services that could be produced in outer space, it is important to note that this paper will deal with the present and potential capabilities of earth-orbiting satellites only. Issues relating to the exploration of the other planets that comprise our solar system's interplanetary space, or of space beyond our solar system, are not considered in this paper. Before considering the nature of the goods and services that could be produced by earth-orbiting satellites, it is instructive to consider the nature of the technical activities that could be performed by earth-orbiting satellites from the perspective of industrial management. These technical activities can be classified broadly as research, development, and operations. In order to define these three phases of technical activity, it is first necessary to define what is meant here by technology. In this context, technology is considered to be knowledge about the physical and life sciences as used for practical purposes (1). The three phases of technical activity may contribute to the development of this knowledge in the following ways:

- <u>Research</u>

    Research may be either fundamental or applied. Fundamental research seeks principles and relationships underlying technology. Applied research demonstrates the potential usefulness of the technology through small-scale experiments.

- <u>Development</u>

    Development reduces the knowledge to practice in a workable prototype form and refines the knowledge for production or commercial exploitation.

- <u>Operations</u>

    Operations is the production, marketing, and distribution of the results of technology.

These three phases of technical activity are applicable to the use of earth-orbiting satellites for the production of goods and services, and efforts aimed at using earth-orbiting satellites for production must proceed through the process of research, development, and operations in a manner similar to earth-based enterprises. The differences between space- and earth-based production become apparent when one considers the nature of the goods and services that could be produced in space and the scale of the investment required.

The goods and services that are (or could be) produced by earth-orbiting satellites may be broadly classified as:

1. <u>Earth Observations</u> - observations of the earth and its environment for the purposes of weather and climate forecasting and the management of earth resources.

2. <u>Communications</u> - the reception and transmission of information from point to point on the earth via a space-based relay.

3. <u>Materials Processing</u> - experimentation with and the manufacturing of high value-to-mass ratio materials.

4. <u>Energy</u> - the collection of solar energy for transmission to the earth as electrical energy.

Some of these goods and services are being produced operationally, at least in part, at the present time, while others are in the research or development phases of activity. Figure 1 is an assessment of the current status of these goods and services. It should be noted that, although a particular product such as Weather Data is shown to be operational, this does not imply that further research and development may not be desirable to improve the value of the product.

Two further general observations are necessary concerning the nature of the goods and services (that could be) produced by earth-orbiting satellites. The first concerns the fact that, at least in the cases of Earth Observations and Communications, the goods and services produced by earth-orbiting satellites are but one element of a much larger system. For example, in the case of weather forecasting, the weather data produced by earth-orbiting satellites provide a synoptic measurement of the state of the atmosphere. The satellite data are combined with data from aircraft, balloons, land stations, and ships and are then used to initialize weather forecasting models. The production of weather forecasts then requires models that represent the physical behavior of the atmosphere and large-scale computational facilities to operate the models (2). Furthermore, a communications system is required to deliver the forecast to the users, since the economic value of a weather forecast is realized only when the user makes a decision based upon the forecast. A similar situation prevails in the case of space communications services where earth stations, terrestrial networks, and switching capabilities are required to integrate the space communications services into a larger communications system. Those space products that have advanced to the development stage also appear to be an integral part of larger systems for data collection, information processing, and communications. It is interesting to note that in those cases where the space-produced goods and services have advanced to the operations phase of activity, the space products are economically superior; that is, they are either of superior quality or cost less than comparable products produced with other technologies.

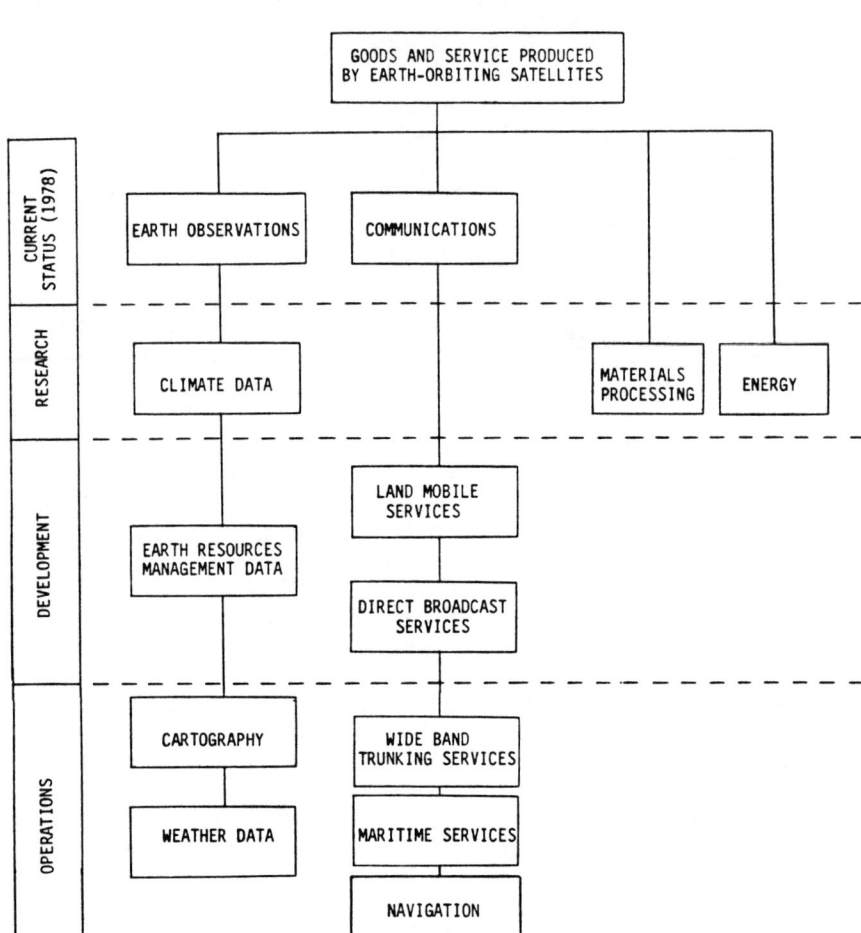

Figure 1. Current status of outer space production.

Transportation

The development of reliable, low-cost transportation has historically been a major factor in economic development. Consider steamships, railroads, automobiles, and aircraft. Each of these forms of transportation has played an important role in establishing both the technical and economic feasibilities of the distribution of new goods and services, as well as in opening new markets. In addition, because of distribution needs transportation itself has become an important sector of the economy.

The initial step toward developing a transportation system appropriate to the production of goods and services by earth-orbiting satellites is the capability represented by the NASA Space Shuttle. A motivating force behind the development of the Space Shuttle was the necessity to reduce the cost of space transportation. Moreover, the Space Shuttle is intended to provide a capability for "routine utility" in earth orbit--economical and routine access to a region that previously has been relatively inaccessible (3). Other capabilities that will eventually be provided by the Space Shuttle and its supporting systems include the retrieval of payloads from orbit for reuse, the servicing or refurbishing of satellites in space, and the deployment and possible fabrication of large structures in orbit. Additionally, the Space Shuttle can serve as an orbiting laboratory for from four to seven crew members for up to 30 days. Advanced power modules now under consideration could extend this orbiting capability to more than ten weeks.

The pricing policy announced by NASA for the Space Shuttle is intended to encourage the full and early use of this new space transportation system by providing a fixed-price contract for transportation, with a 33-month payment schedule. Using a 12-year period of amortization for total operations costs, and prorating the public investment in launch facilities and flight equipment to commercial and non-U.S. users, the shuttle launch costs compare favorably with those of present-day non-reusable launch vehicles. For example, the 1980 commercial launch cost of about $36 million for an Atlas/Centaur payload compares to about $24 million if the same satellite is flown on the shuttle with spinning upper stage. A 1980 commercial launch of a Delta class geosynchronous payload would be about $23 million, compared to about $9 million on a shared shuttle flight with a spinning upper stage (4).

The NASA traffic model used as the basis for establishing the initial user charge structure for the Space Shuttle projects 487 flights during the 12-year period from 1980 through 1992, with a flight projected about once a week during the final six years of that period. Thus, in an average operating year, the revenue from this space transportation system will exceed $950 million in 1975 dollars, not including charges for optional services or amortization of launch and flight equipment. Considering the fact that U.S. commercial and non-U.S. flights make up about 23 percent of this traffic model, this non-U.S. Government use of the space transportation system will generate a cash flow of about $270 million per year in 1975 dollars, including amortization charges (5). Adjusting the gross revenue figure to 1977, the space transportation system would rank about 223 in the list of U.S. industrial corporations, on the basis of sales (6). Compared to U.S. airlines, the space transportation system would rank seventh on the basis of gross revenue (using data provided by Standard and Poor's), between Delta and Northwestern.

Under present institutional arrangements, NASA will be owner and operator of the Space Shuttle transportation system. Other nations will also have launch capabilities during this time period; however, although the Russians probably have the technical capability to develop a recoverable and reusable launch system, as far as we know no other nation is currently developing such a system. Hence it is likely that the NASA space transportation system, once fully operational, will be the most cost-effective means of delivery and recovery of satellites from earth orbit. In this situation, NASA will have a monopoly for space transportation, at least within the United States. The fee structure for use of the space transportation system will be established by NASA, without benefit of a detailed review by a regulatory commission, and with general overview by the Congress. While this monopolistic arrangement may be necessary and desirable during the research, development, and early operational phases of Space Shuttle, it may not be a desirable institutional arrangement for long term. The once comfortable concept of a regulated monopoly is no longer easily accepted in the economics community. Specifically, in the case of NASA, there are at least three problems with using regulation to achieve effective pricing of services. The first is that any commission-established NASA pricing of transportation services will have to deal with the political power of the agency. Studies of powerful regulated monopolies such as utilities indicate that regulation has had little effect on the price levels or structures of these monopolies. The second problem is the tremendous infrastructure required for space activities. The facilities needed to assemble, test, launch, track, control, and recover the Space Shuttle are analogous with the physical plant and distribution system associated with other monopolies such as utilities. The third and perhaps the most important problem is that the regulation process is costly. Since most of the information concerning the cost of operation of the space transportation system would come from NASA, it would be necessary for the regulatory commission to develop a qualified staff to analyze data from NASA and other sources, and for NASA to develop a comparable staff to deal with the commission. With the added delays inherent in a regulatory procedure, and possible further delays in appeals to the courts, the desirability of using a regulatory process is open to question. On the other hand, once the demand for space transportation forecasted by NASA has materialized, it may be desirable from an economic viewpoint to terminate NASA's monopoly (7). This could eventually result in private commercial operation of space transportation systems and vehicles and open the possibility of competition on the basis of price and services.

Future Opportunities for Outer Space Production

Future opportunities for outer space production will most likely come about from those products that are now in the research and development stages of technical activity. The following sections consider some of the economic and policy issues associated with a sample of these products.

Earth Resources Management Data

With a growing worldwide population and an increasing intensity of organized activity in areas such as agriculture, commerce, resource development, and transportation, the need for an improved capability to understand and predict the behavior of the environment is apparent. An ability to measure accurately the state of the environment and the impact of man on the environment is an important step in the development of models which will predict such behavior. These models might also be used to predict levels at which man can hope to extract food and fiber products, as well as other resources, from the earth.

Early experience in the meteorological satellite program indicated that maps of land and ocean features were an interesting by-product of cloud-coverage photographs. It was quickly perceived that images of the surface of the earth in greater detail than is needed for meteorological analysis might provide useful data for management decisions in such areas as agriculture, hydrology, and oil and mineral exploration. The initial step in providing resource managers with more detailed information from space was taken when high-resolution cameras were used to take detailed photographs of the surface of the earth in the early Earth-orbiting Apollo missions. In 1972 LANDSAT-1, the first dedicated satellite capable of providing long-term, repetitive, detailed photography on a global basis, was launched. Since that time, the United States and other nations have maintained an active research and development program in the use of detailed space imagery for resource management. There are many successful users and applications of LANDSAT data, including federal and state agencies and commercial organizations. The U.S. Government is now committed to a continuing program to explore the possible operational uses of data from the LANDSAT satellites. In a recent pronouncement on space policy, President Carter stated that NASA will study the needs for a national LANDSAT-type system, and requested NASA and the Department of Commerce to encourage private investment and participation in this evolving remote sensing system for resource management (8).

The NASA budget for 1980 is expected to set out a six-year, $300-million program to forecast harvests of wheat, barley, rice, soybeans, corn, cotton, and sunflowers--the world's major crops of food and fiber. Principal objectives of the effort will be to improve worldwide crop forecasting and thus gain advance notice of the periodic waves of heavy foreign purchases that have repeatedly shocked the U.S. domestic market for wheat, soybeans, and other crops in recent years. Most of the funds in the new program will support groups in universities and the private sector that will make the detailed analyses of LANDSAT imagery and assess the progress of crops from planting to harvest. This proposed effort is essentially an outgrowth of the three-year Large Area Crop Inventory Experiment (LACIE), in which LANDSAT data were used to predict the production of wheat--the world's most important grain crop--in the United States, Canada, and the Soviet Union. LACIE techniques predicted within one-percent accuracy the Soviet Union's spring wheat harvest of 92 million tons in 1977, while more conventional

Agriculture Department and CIA techniques greatly overestimated the size of the harvest. LACIE also accurately estimated the United States and Canadian winter wheat harvest in 1977; but it was far wide of the mark in estimating the spring wheat crop of the two countries, apparently because much of the latter is grown on small plots in the Great Plains and is difficult to distinguish from adjacent plantings of barley and alfalfa. The new LANDSAT due in 1981, with a 25 percent improvement in ground resolution plus greater infrared sensitivity, should considerably reduce the uncertainty encountered in the Great Plains portion of the LACIE experiment *(9)*.

It is important to realize that the concepts of a resource management system are not new. What is new is that, within the past three decades, the technology of data collection and processing and management-decision sciences have developed to the point where it is now feasible to begin to implement large-scale resource management systems at the national and international levels.

While the details of these resource management systems may differ from application to application, the elements are similar from system to system. Figure 2 schematically illustrates the basic building blocks of such a system. The main elements of the resource management and information system are *(10)*:

(1) <u>Data Collection</u>

> The collection of data concerning the atmosphere, the oceans, and the surface of the earth requires both sensors--to observe or measure the phenomena of interest--and the platforms from which to make the measurements. Some of the platforms, such as ground sensors, ships, and aircraft, operate in or immediately adjacent to the phenomena to be observed. Sensors on these platforms can make direct measurements of the phenomena of interest. On the other hand, earth observation satellites are displaced by distances of hundreds to thousands of miles from the surface of the earth and must remotely sense the characteristics of the phenomena. The development of sensors for *in situ* measurement of environmental phenomena is reasonably mature, while instrumentation for remote sensing is a much newer field that has developed rapidly only since the mid-1950s.

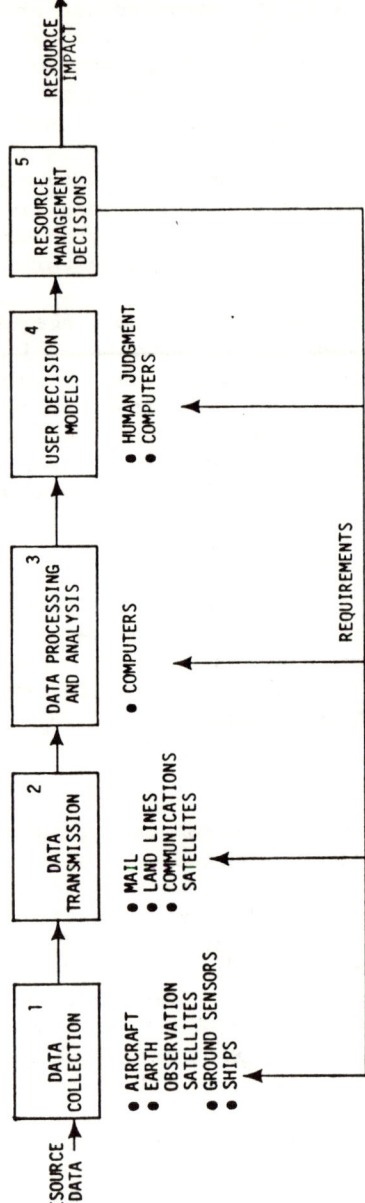

Figure 2. Elements of an information and decision system for resources management.

(2) Data Transmission

The data collected from the many sensors and platforms that are part of a resources management information and decision system must be transported to a centralized facility for processing and analysis. Since much of the data is perishable, timeliness is an important factor. In present systems, much of the information is transmitted from the point of collection to the processing facility using communications satellites, although land lines and undersea cables are also used to move the data.

(3) Data Processing and Analysis

The data collected from the sensors are generally in sensor measurement units; i.e., in the form either of a voltage or of a digital bit stream that is representative of the phenomena observed. At the data processing facility, the data are converted to geophysical units using computer algorithms, combined with data from other sources. At this point, analysis can be performed to describe the state of the enviroment and computer-based numerical models can be used to forecast the behavior of the environment. The large quantity of data collected for assembly into a data base implies the need for large centralized data processing facilities for data processing and analysis.

(4) User Decision Models

At this point in the information chain, the resource manager makes the decision to commit or apply the resources at his disposal. The action and the effects of the decision can result in both costs and benefits. Typical decisions might include:

- establishing the size of the grain storage inventory based upon the estimated harvest yield,

- deciding when to spray or irrigate crops based upon anticipated weather conditions,

- routing a shipment of goods from Seattle to Japan by a southern route as opposed to a northern great circle route to avoid anticipated bad weather that could delay the crossing and possibly damage the ship and its contents.

Since decisions are made with less-than-perfect information, each decision has an associated probability distribution of costs and benefits. User decision models provide a mechanism to investigate and optimize the outcome of the decision. In a user decision model, improved information can be visualized as reducing the risk and uncertainty associated with the decision.

In order to provide input to user decision models, it is usually necessary at this stage to perform further processing that is specific to the needs of the user.

(5) <u>Resource Management Decision</u>

When the decision is made, resources are committed and an outcome is obtained. If the outcome is perceived to be less than optimal, a feedback loop exists from the user to improve the quality or timeliness of the information or the models used to aid in the resource decision process. In this manner, the system is completely driven by user requirements and the user establishes the characteristics of the resource information and management system.

The goods and services that could be produced by earth-orbiting satellites enter in to the Data Collection and Data Transmission elements of this system. The developmental LANDSAT-type satellites, along with the operational meteorological satellites, will play a major role in the collection of the data required for this resource management system. Operational communications satellites will also play a major role in transmitting the wide-bandwidth data collected by the satellite to the point of analysis in the system.

Table 1 presents an overview of the benefits by application and the corresponding system and user costs for an operational LANDSAT system. The benefits and costs are discounted at 10 percent and estimated for an infinite time horizon. The costs are separated into the

the space- and data-management system costs common to all users and the unique user data subsystems. The annual benefits of an operational LANDSAT system are shown in Table 2 (11).

The impact of benefits on various interest groups in the United States is an important concern. If the benefits of an operational LANDSAT program were to accrue exclusively or primarily to a few private

Table 1. Present value of benefits and costs of the LANDSAT operational system.

Systems and Users	Benefit ($ Million)	Cost ($ Million)
Space and Data Management Systems	--	342
Agricultural Crop Information	1,705-3,370	55
Hydrologic Land Use	128	10
Petroleum-Mineral Exploration	202-819	--
Water Resources Management	75-237	--
Forestry	41	122
Land Use Planning-Monitoring	87-278	--
Soil Management	29-52	--
TOTAL (Rounded)	2,260-4,920	530
Benefit:Cost Ratio = 4.3 - 9.3		

Note. Based on Fiscal Year 1976 dollars discounted at 10 percent.

Table 2. Annual benefits of LANDSAT operational system.

Systems and Users	Benefits ($ Million)
Agricultural Crop Information	294-581
Petroleum-Mineral Exploration	64-260
Hydrologic Land Use	22
Water Resources Management	13-41
Forestry	7
Land Use Planning and Monitoring	15-48
Soil Management	5-9
TOTAL	420-968

Note. Based on Fiscal Year 1976 dollars.

corporations, then the investment--however worthwhile--
should be financed by those few who benefit from such
programs. On the other hand, if benefits are widespread
across a variety of groups, interests, and regions, then
programs considered for federal funding have met one
important, necessary condition.

In the case of LANDSAT agricultural crop information,
the main beneficiaries will be farmers and consumers.
Considerable effort has gone into determining more
precisely the exact share by farmers, as against consumers,
but so far the results have not been conclusive.

The main immediate effect of improved crop information
will be more stable prices, while in the longer term
production will increase at lower overall costs per unit
of output. In the economic community there is some
disagreement as to whether stable, nonsubsidized prices
are of greater benefit to farmers or consumers. One
possibility is that in closed economic systems such price
stability as is created by improved information will
mainly benefit consumers in the long run, although in
the short run farmers are likely to be the main beneficiaries. The economic answers on the share in the overall
U.S. benefits depend on many detailed assumptions and
estimates, where a small change in parameters can lead
to a substantial shift in the share claimed for farmers
and consumers--*without changing, however, the total
estimated benefits* accruing to the United States.

Land Mobile Communications Services

Commercial satellite communications systems provide
a fast and reliable worldwide communication link. Most
present-day satellite communication is handled by common
carriers who integrate their space communications into
their existing ground networks. The satellite provides
a long-distance link, with local and regional service
handled by ground link. Since a heavier payload
necessarily increases the cost of the launch, it is to
the economic advantage of the carriers to use a simple,
relatively lightweight satellite, and place the heavy
technical burden on a few large and expensive regional
ground stations.

Such a communication system serves efficiently the
easily accessible, heavily populated areas of the world.
There are, however, many remote, sparsely populated areas
for whom terrestrial communication links are physically
impossible or economically prohibitive. A versatile,

relatively inexpensive satellite communication system is useful for such regions.*

In addition to the advantages remote communities gain from satellite communications services, there is also a large potential market for satellite delivery of social service broadcasts on such subjects as education, medicine, counseling, and public safety.

One system that responds to these needs provides communications between any two points in a given area, with both fixed and moving terminals, on a continuous 24-hour basis. The communications include two-way video service between the stationary terminals and two-way audio service between the mobile terminals. The system includes small, affordable stationary terminals to serve health, education, and other public services provided by federal, state, and local government agencies as well as other organizations; portable terminals for emergencies and disasters; and mobile terminals to serve moving vehicles for emergency medicine, safety, and law enforcement needs.

Such a system requires transferring the burden of the satellite communications network from the earth station to the satellite. This is accomplished by increasing the satellite's transmission power as well as its ability to focus this power. Thus, unlike conventional satellites, which transmit moderately powerful broadcasts over large geographical areas, the public service communications satellites deliver high-power transmission to relatively limited areas. The resultant decrease in the cost and complexity of the receiver stations places satellite broadcasting into the budgetary grasp of the small, independent user and opens the door to a wide variety of social service broadcasts by satellite.

The potential uses are many. Medical students in dispersed geographic locations can receive televised instruction from a major medical center. Consultations can be conducted and data such as EKGs, medical histories, and laboratory reports for diagnostic purposes can be exchanged among physicians at distant hospitals.

---

*In the United States alone, it has been estimated that to reach the last 10 percent of the people would cost as much as reaching the first 90 percent. (Remarks of Casper Weinberger, a news conference on Applications Technology Satellite-F, May 22, 1974, p. 6.)

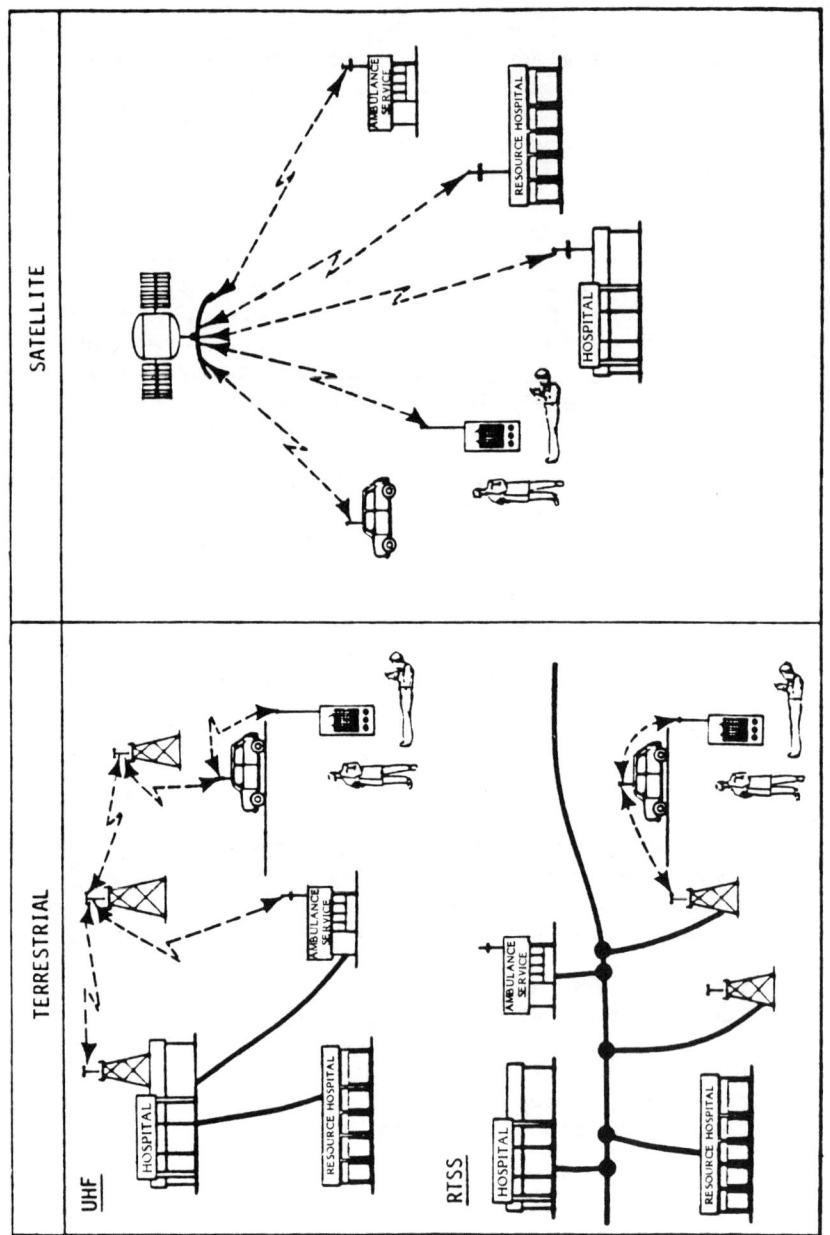

Figure 3. Emergency Medical Service (EMS) system model.

Lectures, as well as audiovisual materials originating at one learning center, can be made available to many geographically dispersed schools, libraries, hospitals, industrial plants, and offices. Schools that are geographically separated but educationally related can, through satellite interconnection, become functional units in special systems, enabling them to share resources and instructional materials designed to meet their common needs.

Libraries and computerized data based across a nation can be linked for instant access and transfer of information. State and interstate communications systems for administration, health, law enforcement, highways, conservation, and public safety can be coordinated. Video teleconferencing can save business companies travel expenses while affording access to scattered offices *(12)*.

The delivery of these and other communications services involving mobile and transportable earth stations have been demonstrated by the NASA Applications Technology Satellites and by the joint Canadian/United States Communications Technology Satellite *(13)*.

During the past two years, NASA has sponsored several studies of the public safety services that could be provided by an advanced communications satellite system. While still in process, these studies give an interesting insight into the potential economic and social benefits and the cost effectiveness of satellite systems for applications such as emergency medical services, fighting forest fires, and federal and state law enforcement agencies.

Satellite-aided emergency medical services is a particularly intriguing application because of the potentially dramatic impact that improved emergency medical services could have on the mortality rate from trauma in the non-metropolitan United States. Figure 3 illustrates the system models used for comparison in a current study of communications in emergency medical services by ECON, Inc., for NASA. This study involved the analysis of emergency medical services systems now in operation in Mississippi, Texas, and West Virginia. These existing systems use terrestrial communications networks to provide voice and data communications between the hospital and paramedics, who can then provide emergency services to the patient under the direction of a physician. For each of the regions studied, we obtained data on the number and distribution of emergencies by trauma type and

CAPITAL COST PER MOBILE UNIT ($)	CONNECTIVITY COST PER CHANNEL MINUTE ($)					
	0.01	0.10	0.50	1.00	2.00	5.00
2,500	23.57 / 50 / 1.000	22.21 / 50 / 1.000	16.20 / 49 / .978	8.84 / 49 / .978	1.33 / 8 / .054	.22 / 1 / .002
5,000	17.42 / 50 / 1.000	16.06 / 50 / 1.000	10.09 / 49 / .978	2.84 / 38 / .728	.86 / 7 / .042	.21 / 1 / .002
10,000	5.13 / 50 / 1.000	3.78 / 49 / .978	1.13 / 11 / .091	.74 / 6 / .034	.52 / 2 / .004	.17 / 1 / .002
15,000	.74 / 6 / .038	.71 / 5 / .024	.58 / 4 / .015	.53 / 2 / .004	.46 / 2 / .004	.14 / 1 / .002
20,000	.53 / 2 / .004	.52 / 2 / .004	.50 / 2 / .004	.47 / 2 / .004	.40 / 2 / .004	.11 / 1 / .002
50,000	.26 / 1 / .002	.25 / 1 / .002	.24 / 1 / .002	.22 / 1 / .002	.19 / 1 / .002	.00 / 1 / .002

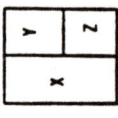

X = Millions of dollars saved per year.

Y = Number of states for which a satellite communications system will be cost effective at the given capital and connectivity costs.

Z = Portion of total channel minutes required for EMS in the United States that is required by the Y states that would find the satellite system cost effective at the given cost levels.

Figure 4. Cost effectiveness of satellite communications. Shown are cost savings (in millions of dollars per year) of a satellite system over a UHF system in which the upper mileage limit of radio coverage is attained and the Emergency Medical Service owns 50 percent of the fixed ground equipment.

the communications traffic associated with these emergencies. From the participating physicians, we obtained estimates on the reduction in mortality rates for each trauma type attributable to voice and data communications between the paramedic and the physician. Using data on population density and terrain type by state as the basis for generalization to the non-metropolitan regions of the United States, we have concluded that about 59,000 lives per year could be saved with an emergency medical communications system covering the entire non-metropolitan United States. Using conservative estimates for the values of the lives saved, and omitting further benefits that might be possible through reductions in morbidity as a result of better emergency medical attention, we estimate the economic benefit of this reduction in mortality to be approximately $2 billion per year. The question of the cost effectiveness of satellite systems to provide these services is a function of the designs of the competing terrestrial and satellite systems. Since the satellite system to provide these services has not been designed, it is necessary to analyze cost effectiveness in a parametric fashion. Using existing terrestrial systems as a reference point for comparison, the cost effectiveness of satellite-aided emergency medical communications is shown in Figure 4 as a function of the connectivity cost and the capital cost of the mobile unit. On the basis of this analysis, we have concluded that satellite-aided emergency medical communications could be cost effective over a very wide range of connectivity and equipment costs.

Both technical and policy issues are involved in the implementation of satellite-aided emergency medical services. Although NASA has demonstrated the capability to communicate with both moving vehicles and small, portable terminals especially designed for medical use, the in-orbit switching capability needed to serve many users simultaneously has not been developed and the frequencies for this type of service have not been assigned. In addition to the technology, market aggregation is also a major issue. As opposed to the relatively monolithic trunking communications market, the emergency medical services communications market is quite diffuse. Emergency medical service districts are organized at the county level and there may be many such districts in a given state. Moreover, since it is not possible to establish a consumption-related pricing mechanism, the funding for emergency medical services is obtained from general taxation rather than from fees. This moves the decision to implement these systems from the market to

the political and policy arenas. The President recognized
these issues in his space policy announcement, and directed
the National Telecommunications and Information Administration to assist in market aggregation, technology transfer,
and possible development of public satellite services in
areas such as health services that receive little attention
from commercial satellite operations (8). NASA and other
federal agencies are currently proposing to continue
the development of satellite-aided emergency medical
services through a demonstration project involving medical
support to the offshore oil industry in the Gulf of Mexico.

Research Product Areas

The development of the capabilities to manufacture
specialized materials and to collect solar energy for
retransmission to earth as electrical energy are typical
of goods and services that are now in the research phase
of technical activity. Both of these product areas
require further research and development to demonstrate
technical and economic feasibility, but are especially
interesting insofar as they could lead to an eventual
large-scale industrial activity in the space environment.

Space Materials Processing. The use of the natural
space environment as a laboratory in which to perform
materials research has long been of interest to space
scientists. Small-scale materials processing experiments
began in the later Apollo missions in 1971 and continued
through the Apollo-Soyuz flight in 1975. Since that
time, short-duration experiments have continued using
unmanned rockets and drop test facilities. NASA plans
to begin longer duration experiments again in 1981 using
the Space Shuttle. NASA recently selected a set of
experiments to fly in the early Space Shuttle missions.
These experiments include work with semiconductor devices,
crystals, glasses, monodisperse latexes, and human blood
cells.

In space processing, a material is transported into
space. A value-added operation is then performed on that
material and the resultant product is then returned to
earth. In some cases, the result of a space materials
processing experiment may be information that can be
used to improve a process that is performed on earth;
however, the most likely result of operational space
materials processing will be a physical product that can
be returned to earth for subsequent sale.

A recent study of space materials processing by the National Academy of Sciences concluded that the work performed to date has not shown any examples of economically justifiable processes for producing materials in space, and urged NASA to concentrate on the scientific and technical opportunities of future experiments. The review suggested that, in the near term, NASA should concentrate on a development and demonstration phase of activity. In this phase, NASA would perform experiments to define the potentials and limitations of materials experiments in space. If this phase is successful in demonstrating a role for space materials processing, the review suggested that the Space Shuttle and its supporting payload equipments be developed as a national space laboratory for materials processing experiments. Possible commercial activities in space materials processing are viewed as following these earlier phases of activity. In its study, the National Academy of Sciences stresses the fact that the national laboratory concept should be pursued only if the demand for this facility exists and if prospective experimenters are willing to pay for time on the facility (14). In recognition of the long-range aspects of this program, NASA has embarked on a series of studies to define the needs of a space laboratory facility and to investigate some of the market and institutional questions that will arise as commercial products are proposed. One version of the proposed NASA program is shown in Figure 5. The reader should realize that activities shown beyond the early flights of the Space Shuttle represent current planning, but that actual events will be determined to a great extent by budget actions.

In addition to technical issues, some of the important economic and policy questions relating to the national laboratory concept and the potential commercialization of products developed in this facility are:

- What are the expected technical options and costs (1) to NASA (2) to industry?

- What is the anticipated user demand for the facility as a function of capability and user cost?

- What is an appropriate schedule of user costs?

- How will the facility be operated and managed?

- Could the facility be a private sector venture?

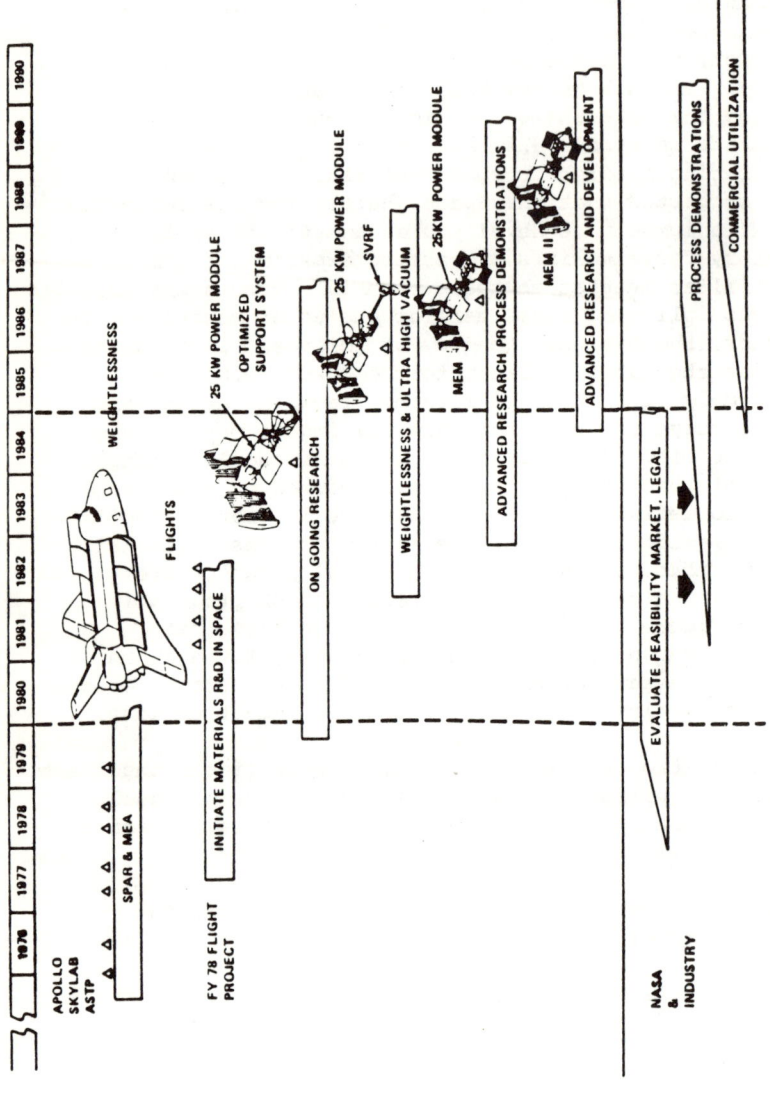

Figure 5. Materials processing in space -- long-range plan.

- What is the policy for data rights and patents for joint endeavors between NASA and industry?

Energy Systems. The possibility of generating large amounts of electrical power in space, converting the power from DC electric to RF microwave, and transmitting it back to Earth for conversion back to DC for distribution was proposed about ten years ago by Dr. Peter F. Glaser. At that time, the idea seemed to offer only a remote hope for practical implementation and was difficult to conceive as being an "economical" idea to pursue. Advancing technology and energy costs have combined to make the satellite solar power system (SSPS) an increasingly interesting concept.

Figure 6 schematically illustrates one concept of an SSPS that has been subjected to study. This concept appears to provide a technically feasible long-term energy source, independent of hydrocarbon fuels, that could be integrated into existing electrical energy distribution networks. While the basic physics of these systems are understood, many technical uncertainties must be resolved and new technical capabilities developed before this concept can be demonstrated. Foremost among these technical factors are questions pertaining to the transmission of large quantities of RF energy through the atmosphere, the ability to fabricate, assemble, and maintain large structures in space, and the need for a transportation system with greater capability than the Space Shuttle.

The SSPS project is unusual in several ways. First, its sole purpose is to supplant an existing capability--baseload electric power generation--in contrast to virtually all other space projects, which provide a substantially new or improved capability. As such, the SSPS must compete economically in the environment of the baseload electric power generation system as it will exist in the 2000 to 2050 time period, when SSPS could be implemented. The baseload system of the period is likely to consist of current "conventional" generating plants, coal and nuclear, and may include new types of plants, for example, fusion. Thus, there are two ways in which SSPS could fail to achieve an economic status from which its implementation would be desirable: It could fail to be cost-effective compared to conventional plants, or there could be major breakthroughs in non-conventional plants, resulting in a substantial decrease in the cost of generating electrical energy.

The second unique aspect of SSPS is its state of development. Unlike fusion, the physical principles at work in the SSPS are well understood and the concept

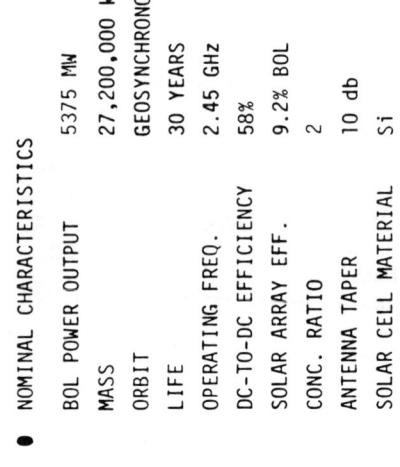
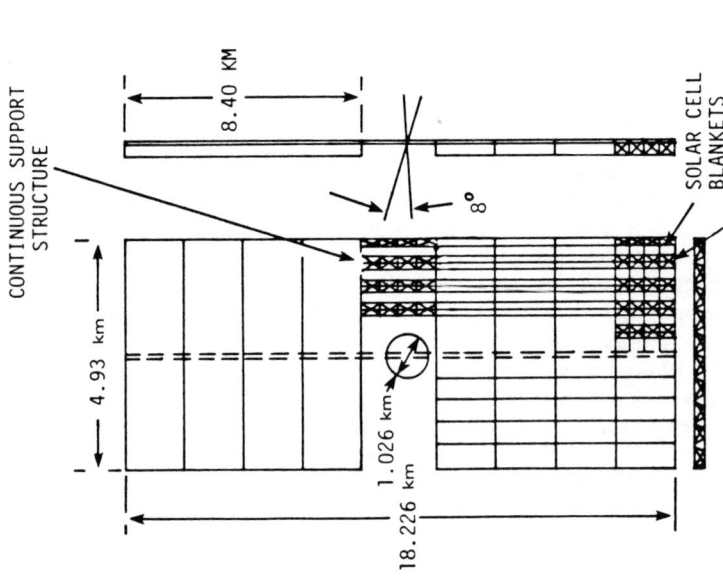

Figure 6. Satellite solar power system (SSPS) configuration.

has been essentially demonstrated on a very small scale. There remain, however, major technology developments to be accomplished to obtain the necessary cost goals. These developments can only be accomplished over a period of several years and at considerable expense. Furthermore, there is considerable uncertainty in the outcome of the necessary development efforts.

Third, both because of the time required for technology development and because of the adequacy of conventional energy sources today, SSPS will not be implemented for at least 20 years. Thus, the payback period on SSPS research and development expenditures will be quite long--measured in decades. Combined with its strictly civilian nature, this places SSPS in a position shared, perhaps, only by fusion.

Since the fundamental purpose of SSPS is an economic one, it is appropriate to evaluate the concept in an economic context. To do this requires estimating SSPS costs, despite the uncertainties that exist today. Cost data can be used as a basis for two types of decisions: first, the decision to pursue the SSPS concept and at what level of effort, and, second, if the SSPS concept is to be pursued, what areas should the research and development efforts focus on *(15)*.

Given today's state of knowledge concerning the technical options involved in the SSPS concept, the economics of SSPS are highly uncertain in terms of what the system will cost and what the system can cost. The development time of this exciting concept is measured in decades and the costs in tens of billions of dollars. At the present time, the full-scale development of an SSPS cannot be justified on economic grounds. However, as in the case of other long-term energy prospects, it is important to preserve our option for this decision at a future date. Since a great deal of the technology needed for an SSPS is also applicable to other projected space applications, a broad-based technology program focused on resolving the major technical and economic uncertainties that surround the program is probably the best approach at this time.

Other Economic and Policy Issues

Many students of space policy, in both the public and private sectors, have suggested that the development of the goods and services described in this paper has been inhibited by a scarcity of funds. In fiscal year

1979, NASA will spend about $4.3 billion on all civilian aeronautics and space activities. Within this amount, less than 10 percent of the NASA budget is allocated to funding the systems to produce these goods and services. To some extent, this lack of funds in these areas of practical application is probably attributable to the inability of NASA to place these opportunities for outer space production in the context of everyday national economic concerns, needs, and requirements (16). This may, in part, be attributable to the interpretation of the Space Act of 1958, accepted by NASA, which emphasizes the role of the agency in research and development, not as an operator of systems. This situation could be corrected through legislation introduced by Senators Stevenson and Schmitt in the 95th Congress; however, it is not clear whether that legislation is supported by the executive branch and, without strong support from the administration, its fate is in question (17). Since the present administration has not proposed to take a strong initiative in the funding of systems for production in outer space, it is of interest to explore other possible sources of funds.

One possible source of additional funds is through cooperative efforts with other nations. Some of the national competitive edge of the space program has worn off in the past ten years and the federal government is to be commended for the initiative shown in the area of international cooperation. However, the prospect of obtaining a significant influx of funds for systems to produce goods and services is complicated by issues of national ownership of the systems and the products that they turn out. This leads to the question of the role of private enterprise in the development and operation of these systems.

Because of the historic role of the federal government in the provision of space technology, the private sector has developed a system of dependence on the government for both the systems and the products of space enterprise. Recent initiatives by both the President and Congress appear to recognize the limitations of this situation and have called for increasing participation by the private sector in those earth observation and communications systems that could be transitioned from development status to operations in the near term (3,8). While these initiatives will probably have a positive effect on these near-term opportunities, what is really required is a more far-reaching examination of the role that could be played by the private sector in the entire space program. The role of the federal government in research and development is not questioned. However, the role of the federal government as the sole operator of space transportation systems and systems to produce practical goods and services may be inhibiting and should be scrutinized, with a view to the introduction of the entrepreneur into outer space production. This lack of entrepreneurship is probably the weakest link in the United States space program today.

References and Notes

1. J.B. Quinn and J.A. Mueller, "Transferring Research Results to Operations," *Harvard Bus. Rev.*, January-February 1963.

2. P.M. Wolff and P.J. Kesel, "Accurate 24-Hour Weather Forecasts -- An Impending Scientific Breakthrough," in *Earth Observation Systems for Resource Management and Environmental Control*, D.J. Clough and L.W. Morley, Eds., NATO Conference Series II: Systems Science (Plenum Press, New York, 1977).

3. Bill S.3530, Congressional Record, Vol. 124, No. 153, 27 September 1978, pp. 1-4.

4. J.F. Yardley, "The Space Transportation System, A Public Investment - A Public Utility," *Bus. Week*, September 25, 1978.

5. J.M. Smith, R.F. Kranupf, and L.J. Jacobson, "Development and Implementation of NASA's STS Reimbursement Policy," *J. Contemp. Bus.* 7 3 (1978).

6. The Fortune Directory of the Largest U.S. Industrial Corporations, *Fortune*, May 8, 1978.

7. W.A. Jordan, "Commercial STS Prices: An Economic Analysis," *J. Contemp Bus.* 7 3 (1978).

8. U.S. Civil Space Policy, *The White House Fact Sheet* (Office of the White House Press Secretary, Washington, D.C., 11 October 1978).

9. H. Simmons, "Carter Space Policy Great for Farmers," *Astronautics & Aeronautics*, January 1979, pp. 9-12.

10. B.P. Miller, "Commerce Ventures That Exploit the Observation of Earth from Space," Paper presented at the Conference on Space Commerce: New Options for Economic Growth, New York, November 1978.

11. C. Buffalano et. al., *A Cost-Benefit Evaluation of the LANDSAT Follow-On Operational System* (Document No. X-903-77-49, NASA Goddard SFC, 1977).

12. Science Policy Research Division - Library of Congress, *World-Wide Space Activities*, Report Prepared for the Subcommittee on Space Science and Applications of the U.S. Congress Committee on Science and Technology (Pub. No. 85-503, Government Printing Office, 1976).

13. NASA, *Communication Technology Satellite* (CTS File No. 3100-27, NASA Lewis Research Center, 1976).

14. Space Applications Board, "Materials Processing in Space", in *Report of the Committee on Scientific and Technological Aspects of Materials Processing in Space* (National Academy of Sciences, Washington, D.C., 1978).

15. G.A. Hazelrigg, Jr., "Costing the Satellite Solar Power System," Paper presented at The 25th Anniversary Conference of the American Astronautical Society, November 1978 (Paper No. AAS-78-166).

16. K.P. Heiss, "Economic Opportunities in Space Enterprise," *J. Contemp. Bus.* 7 3 (1978).

17. See, for example, S.3625 - Earth Resources Information Satellite Act of 1979; and S.3530 - Space Policy Act of 1978.